Ergonomics Laboratory Exercises

Ergonomics Laboratory Exercises

Timothy Joseph Gallwey
Leonard William O'Sullivan

CRC Press is an imprint of the
Taylor & Francis Group, an **informa** business

Cover: An example of Leonardo da Vinci's flying machine made to his drawings by the Science Museum, London, and donated to the University of Limerick by Dr. Tony Ryan of Guinness Peat Aviation. It hangs in the Atrium of the Foundation Building and raises such ergonomics issues as: body dimensions, layout of controls, perception and vision, learning, physiology, and the ability to perform physical work.

CRC Press
Taylor & Francis Group
6000 Broken Sound Parkway NW, Suite 300
Boca Raton, FL 33487-2742

Printed in the United States of America on acid-free paper
10 9 8 7 6 5 4 3 2 1

International Standard Book Number-13: 978-1-4200-6736-1 (0)

Library of Congress Cataloging-in-Publication Data

Gallwey, Timothy J.
Ergonomics laboratory exercises / Timothy Joseph Gallwey, Leonard W. O'Sullivan.
p. cm.
Includes bibliographical references and index.
ISBN 978-1-4200-6736-1 (alk. paper)
1. Human engineering--Problems, exercises, etc. I. O'Sullivan, Leonard W. II. Title.

T59.7.G355 2008
620.8'2076--dc22 2008019819

Visit the Taylor & Francis Web site at
http://www.taylorandfrancis.com

and the CRC Press Web site at
http://www.crcpress.com

Contents

Preface

This book provides a series of exercises for the laboratory work aspect of the formation of professional ergonomists as evaluated by CREE (Centre for Registration of European Ergonomists) according to HETPEP (Harmonising Education and Training Programmes for Ergonomics Professionals). CREE evaluates applicants from European Union (EU) ergonomics societies to register as a European Ergonomist (EurErg) according to HETPEP criteria, and to demonstrate their professional competence, in order to facilitate movement within the countries of the EU.

HETPEP specifies that the education component must be supplemented by "Laboratory exercises, [which] are in addition to the ... [classroom] hours and are an integral component from the beginning of the education period ... . [They] should prepare the student for later training and experience ... [and] should comprise approximately 30 to 35 day-parts (3 hours each) that should total about 100 hours during the academic period." But the concept of "laboratory work" in ergonomics appears to be confused, or is at least unclear, for significant numbers of people.

Ergonomics is not a pure science like physics, chemistry, or experimental psychology. Ergonomics knowledge is not sought for the fundamental purpose of understanding how something works, what its laws are, or which theories are the most valid. That is the pursuit of science and, although ergonomics has a science basis, the essence of the profession is to use the findings of science to solve problems in the here and now. It is an applied science like engineering. Stokes (1997) has illustrated this point with a two-dimensional array of fundamental understanding versus consideration of use, where high understanding and low consideration of use is labelled as the Bohr quadrant, and the reverse is the Edison quadrant. He points out how high understanding from basic research led to solid state devices but now high considerations of use require the improvement of the performance of these devices which in turn requires further basic research. Likewise, ergonomists need to combine high understanding of the scientific basics with high considerations of use or application, which he classifies as Pasteur's quadrant. That is where we must aim.

Ergonomics cannot be learned out of a book. It must be learned by doing, with an applications-oriented ethos. It requires "hands-on" learning, where students see major aspects of relevant scientific phenomena for themselves, gain experience in how to collect data on them, and learn how to apply them. It is well known that active learning is much more effective than passive learning (e.g., see Czaja and Drury, 1981). Traditionally, this point has been demonstrated in office work and factory jobs, but it also applies in academia. Laboratory work in its general sense is the best way for students to receive active learning in an academic course. But they also need to learn about the safeguards required for obtaining valid and reliable data, they need to learn how to interpret its meaning, and they need to learn how to devise solutions to real world problems.

To some people "laboratory work" appears to be synonymous with people in white coats, using sophisticated and expensive high-tech equipment to collect highly accurate data on complex activities under tightly controlled conditions. That is one of the meanings used here, but it also includes what is sometimes called practical work, or practicals, or praxis, or *Ubung* in German, or practicum elsewhere, that is, practice or practical exercises. But these wider terms are also used at times in other contexts to refer to gaining general practical experience of working in industry, so their usage could lead to confusion. At the risk of telling people what they know already, it is first of all necessary to establish the essential nature of the type of work that is expected to be undertaken. We define "laboratory work" as an investigation with the following characteristics:

- An activity is performed to achieve a specific end (e.g., assemble parts).
- The performance is observed in a scientific manner (i.e., with some controls).
- Data are collected on that performance (probably with "scientific instruments").
- The data are analysed by scientific methods (e.g., mathematical statistics).
- The results are compared with those published in the scientific literature.
- A scientific report is drawn up that gives conclusions and recommendations.

Such investigations include the traditional sophisticated laboratory work but, in this document, we also include simpler investigations such as stopwatch studies, and pencil-and-paper exercises. The bulk of the work is likely to be performed on campus, with data collected on tasks performed by the course participants, in the classroom or the laboratory, under conditions not as tightly controlled as in proper research work. But, provided the characteristics of the previous paragraph apply, they will satisfy our definition.

Students should experience the inherent variability of the data collection process, and learn how to limit the amount of variability in their data by good experimental design and practice. They should have to write scientific reports on the work in order to obtain first-hand exposure to the process of analysis, comparison, inference, deduction, and drawing of conclusions. It should include consulting current scientific journals to ensure exposure to the latest findings, and it should usually also require the use of sophisticated statistical analysis.

Classroom work should be supplemented by field type studies. Preferably the students should investigate real work sites, with real employees, who perform real tasks, in real jobs, in a nonacademic establishment. But it is also possible to perform such studies on campus. They should expose students to the difficulties of doing such work, explore techniques for getting reliable and repeatable data at such sites, and develop skills in dealing with people in the workplace. The work should include data collection and analysis followed by a formal written report similar to those required for the other investigations. It will usually be done on a teamwork basis so it will enhance the skills needed for working in a team.

For some subject matter seminars, essays, tutorials, or self-work assignments are more appropriate mechanisms to support the lecture material. Hence, some material is provided to meet these needs. While the emphasis is on the relevance of the material to real world issues of ergonomics, the stress must be on understanding the fundamental principles involved and how they relate to relevant theoretical issues.

Part of the aim is to take a holistic approach so as to develop a systems view. But the laboratory work is not just an academic exercise. In many situations the state of ergonomics knowledge is insufficient to provide the basis for acceptable and sound solutions. So practitioners need to be able to collect their own data in an accurate and reliable fashion. That needs a good grounding in laboratory work.

Finally, the process helps to sharpen the students' critical reading of the scientific literature. It should help them to tease out reasons for differences in results, and to deduce appropriate measures to adapt reported results to their needs. It should help them to select the most appropriate methods and results in devising their solutions to the problems they address. It should also engender in them a respect and a desire for scientific rigour.

Czaja, S.J. and Drury, C.G., 1981, Aging and pretraining in industrial inspection, *Human Factors*, 23, 485–494.

Stokes, D.E., 1997, *Pasteur's Quadrant: Basic Science and Technological Innovation*, Brookings Institution Press, Washington, D.C.

The contents of this book are supported by additional material which is available from the CRC Web site: www.crcpress.com. This includes a list of possible equipment vendors.

Under the menu Electronic Products (located on the left side of the screen), click on Downloads & Updates. A list of books in alphabetical order with Web downloads will appear. Locate this book by a search, or scroll down to it. After clicking on the book title, a brief summary of the book will appear. Go to the bottom of this screen and click on the hyperlinked "Download" that is in a zip file.

Or you can go directly to the Web download site, which is http://www.crcpress.com/e_products/downloads/download.asp?cat_no=67362.

Acknowledgments

Some of the exercises detailed in this document owe a debt to the former ergonomics staff of the University of Birmingham (especially Dr. Ben Davies and Prof. Nigel Corlett), and a heavy debt to the State University of New York at Buffalo (Prof. Colin Drury). Their contributions are gratefully acknowledged and hopefully these are well referenced, but inevitably there are probably some topics and ideas that have not been referenced sufficiently. We ask for their forbearance. However, all errors and deficiencies are the responsibility of the authors.

Many students have contributed ideas and improvements over the years to the work detailed in these experiments. Important contributions have come from specific graduate assistants in the universities in Buffalo, Windsor (Ontario), and Limerick. But a significant contribution has also been made by all course students through demonstrating deficiencies in wording and procedures, and these have been extremely valuable in refining the work.

Considerable contributions have come from authors such as Sanders and McCormick, Mundel, Barnes, and Konz and these have been acknowledged wherever we have been conscious of it. However, there are probably instances where their ideas and contributions, and those of others, have been absorbed to the extent that we are no longer conscious of them, and we hope for their forbearance and that of their publishers.

We offer a special thanks to the publishing staff — Richard Steele in the UK, Cindy Carelli, Amber Donley, and Gail Renard in the US — for their help in all the activities of producing a book.

We owe a particular debt to John Collins, who did the drawings that supplement this text to assist users in carrying out the exercises detailed here.

About the Authors

Dr Tim Gallwey has degrees in engineering and ergonomics. After 10 years in heavy engineering and the automobile industry, he moved to academia in Canada and then Ireland where he established a master's degree program in ergonomics at the University of Limerick and started the Irish Ergonomics Society. He was actively involved in the creation of the Centre for the Registration of European Ergonomists and led its first course accreditation. His research is mainly in musculoskeletal disorders, and he has been a partner in EU projects in this area. Currently, he is an independent ergonomics consultant.

Leonard O'Sullivan has a bachelor's degree in materials and construction and a master's and Ph.D. in ergonomics. He joined the University of Limerick as a lecturer in the Department of Manufacturing and Operations Engineering in 2003, where he lectures in industrial and product design ergonomics. His main interests are in musculoskeletal disorders, and he has been involved in several European Commission funded ergonomics research projects.

1 Report Writing

Important parts of the academic process, and the development of a deep understanding of ergonomics, are the analysis of data and the writing of scientific reports. The process of reading and dissecting various sources of information, breaking them up into distinct pieces, and reassembling them into a series of sound, cohesive points is one of the most challenging intellectual tasks. It is also one of the best means of developing a good understanding of the material and of helping students to learn how to organise data to make a scientific case or argument. For these reasons, the whole process of laboratory work would be incomplete without having to write up the work in a professional, scientific manner; therefore, it is an important part of these exercises.

Obviously, the requirements and styles of employers differ from each other and from those of scientific journals, just as the requirements of scientific journals differ, and hence it is impossible to provide guidance on what is needed by each. However there are general issues that have to be addressed, and these have been incorporated into a set of requirements for laboratory reports. In some cases, the work may not warrant a lengthy report, especially where a deep theoretical issue is not being examined; some employers actually prefer a shorter, more succinct style. For this reason three report styles are presented. Accompanying the short report-style document is a sheet for comments that can be ringed where appropriate to indicate particular shortcomings. Students, especially in the early stages of such studies, often have only a hazy idea of what is expected of them. To clarify this aspect, marking schemes with questions and pointers are discussed as well. To ensure that correct scientific notation is used, and correct formatting of the document, additional detail is provided on report presentation.

The HETPEP (Harmonising Education and Training Programmes for Ergonomics Professionals) document requires a final piece of project work to integrate study material and to provide particular depth in one area; this has its own special writing-up requirements. Writing laboratory reports in the style provided gives the students good training in how to construct such a final document, but a separate guide is provided for the style and structure of the project or thesis report. The requirements of different institutions will probably not be the same as that given here, but the general form of these documents has been developed over several years and should therefore match most of the requirements of most employers and publications.

Because ergonomics students come from a variety of scientific and engineering backgrounds, they may be accustomed to different conventions and report structures. This may be particularly apparent in different approaches to notation and the designation of units. The conventions used here are those largely accepted by most journals devoted to ergonomics, and employ SI units.

The analysis aspects of these experiments require a good ability with statistics, more than that obtained in an introductory course. Students should have had exposure to a thorough course in the more advanced aspects of the design of experiments. In particular, an understanding is needed of the issues associated with mixed-model designs and expected mean squares in the analysis of variance (ANOVA), and the process of transforming variables to meet the normality requirement of ANOVA. Around 40 hours of lectures is expected. The importance of having this background becomes particularly noticeable in project/thesis work. Serious problems can arise with both design and analysis if these areas are not well understood and implemented. Hence, many of the experiments incorporate particular emphases on these issues. The authors have worked with SPSS (Statistical Package for the Social Sciences; www.SPSS.com), as reflected in some of the material, but any one of the well known computer packages will be suitable.

1.1 SHORT LABORATORY REPORTS

Report writing is an important part of any job and a difficult discipline that needs to be learned and practiced. Short reports are limited to a maximum of six pages, and should be carried out according to the following format, incorporating also the requirements specified in Section 1.3, Report Presentation.

First Page (10% of Marks)

Concept Examined: This constitutes the top half. It is not a description or summary of what was done, but rather a brief expose or essay on the underlying theme or concept studied in your labwork. Certain aspects of ergonomics relate directly to the topic of the lab and so provide the basis for the work performed. Describe them. The treatment must be conceptual and general and end in a sentence stating the concept, principle, or idea examined.

Method: The bottom half of the page is this section. It must give sufficient detail that someone else will be able to repeat what was done. Minutiae of benches, etc., for example, are not relevant but information on apparatus used, procedure, and type of person used is relevant, provided that they may have affected the results achieved. It will help you or others if or when the work has to be repeated.

Second Page (15% of Marks)

Results: The top one third or so is this section. It must describe in words what information came to light from the work. Do not try to explain it or refute its validity, etc., here. Just state in words and with some data what was found, especially findings that run counter to expectations.

Discussion: The middle third of the page is this section. This is not a rehash or summary. Here, consider the quality of the experimental work, its validity, possible reasons for unexpected results, and what could have been done differently.

Conclusions: This is the last third of the page. It must consist of a series of numbered one-line or two-line statements of what the work revealed. Do not leave the reader wondering what to make of it all. The statements must relate to what was done, and must be supported by the data. Do not indulge in general, unsubstantiated speculation.

Last Four Pages (35% of Marks)

A set of appendices is given here. These consist of such items as tables of data collected, graphs, process charts or flow diagrams, sketch of the workplace, sample calculations, etc.

Writing Style (20% of Marks)

Correct grammar, spelling, and sentence construction must be used. What was done must be written in the past tense; by the time the report is written, it will be history. Avoid padding, waffling, and irrelevant statements. It must be written in the third person, that is, do not use "I" or "we" or "you" but rather "It was found that …". Telegraphic or military style is also not acceptable such as "Timed by stopwatch". Sentences must contain a finite, transitive verb.

Understanding of Concepts (20% of Marks)

The report must demonstrate that the student(s) understands the concept(s) involved, the techniques used, the meaning and relevance of the results obtained, and their implications.

1.2 SHORT LAB REPORT GRADING COMMENTS

Student(s)____Lab Group_____
Lab Topic____

Concept examined: waffle, summary, too short, too long, extraneous info, something other than concept, does not stand alone.

Method: summary, lacks detail, waffle, not a method, describes the wrong thing, contains material of other sections, incomplete.

Results: data not presented, discussion, incomplete, missing, what was achieved?, waffle, method/procedure, complaints, says nothing, merely refers elsewhere, data table put here, little or no link to the concept(s), not described verbally, wrongly stated.

Discussion: results, summary, rehash of results, points missed out, waffle, complaints, missed out altogether, method, little relevance.

Conclusions: not numbered statements, incomplete, missing, not related to the data, what did you get?, waffle, complaints, missed out, not justified by data, summary, wrong, against the data.

Tables: badly drawn, wrong labelling, units omitted, data omitted, one or more not included, values wrong or wrong data, not labelled.

Figures: wrongly drawn, wrongly labelled, info omitted, axes wrong, one or more not included, legend omitted, units wrong or omitted, label omitted, dimensioning problems.

Format: units wrong or not SI, pages wrong way round, report structure wrong, work asked for not done, too long, too short, headings wrong, wrong page order, not printed on one side of page only, not typed or printed, Discussion or Results text in appendices, correct sheets not used, sheets submitted not originals.

Writing: bad spelling, errors not corrected, bad grammar, faulty punctuation, wrong tenses, non-sentences, sentences not flowing, mixed singular and plural, telegraphic, not third person, use of abbreviations, padding, unclear.

Understanding: inadequate, wrong, missed the point, unclear.

Sample calculations: omitted, wrong, incomplete, too extensive, too brief.

Critique: omitted, wrong, incomplete, sketchy, on the wrong topic, improvements not outlined, did not need the lab to show what has been presented.

1.3 REPORT PRESENTATION

Cover sheet: This consists of a declaration sheet that it is the author's own work; sources are fully acknowledged and signed (by all, in the case of group work) prior to submission.

Paper and usage: The report must be on A4 paper or similar size, using one side only, with 2 cm margins.

Text: must be typed or printed in 12 point.

Orientation of sheets: Normally it will be "portrait", however, for some tables and figures, it will have to be turned through 90 degrees (i.e., "landscape"). In the latter case the bottom of the table or figure must lie by the right hand edge of the report when it is laid out open on the desk.

Tables: Must be labelled descriptively across the top (e.g., "**Table XX.** Times for operators to perform each combination of conditions"). Rows and columns must be labelled for the variable represented and (in brackets) the unit of measure. *Note:* the quotation marks indicate the exact type of wording to be used but must not be included in your report.

Figures: Consist of all diagrams, graphs, charts, pictures, photographs, drawings, sketches, etc., and they must be labelled across the bottom (e.g., "**Figure YY.** Mean time of each group for each condition"). Axes of graphs must be labelled for the variable represented with (in brackets, see below) the unit of measure. Where a graph has more than one set of points there must be a legend to identify each set, and it is still called a "figure" and not referred to as a "graph".

Plotting: Individual lines on a graph must be labelled separately or the plotted points differentially identified (e.g., by using circles for one, triangles for another, and so on with a legend to identify each). Use metric (1, 5, 10 mm) paper or a computer package such as Microsoft EXCEL. The independent variable (i.e., what is altered deliberately [e.g., task difficulty]) must be along the horizontal axis and the dependent variable (i.e., what is measured for the results; e.g., time) must be along the vertical axis.

Units and notation: Must always conform to the Systeme Internationale (SI) and no others may appear in the report (e.g., masses must be in kilograms [kg], weights and forces in Newtons [N], lengths in metres [m] and millimetres [mm], velocities in metres per second [m/s], and times in seconds [s] usually but also in centiminutes [cmin], minutes [min] and hours [h] on occasions, but do not mix them such as seconds and minutes for the same quantity).

Binding: Must be such that pages can be read easily without having to dismantle the report. Where larger sheets are used (e.g., A3) bind the left-hand edge and fold in from the right to clear the binding, or staple the top left-hand corner, fold at the centre of the page, and then fold back or clip off the top right-hand corner.

Format: Ensure that the text is justified both sides, leave a blank line between paragraphs, use single spacing for lab reports, and one-and-a-half or double spacing for projects and theses.

1.4 LONG LABORATORY REPORTS

Construct the report as specified below, written in your own words. Mode of presentation and marking requirements are defined separately. As a general guide, see *Ergonomics* or *Human Factors* journals.

1. Introduction (15% of Marks)

It must review previous work in a critical fashion (i.e., main findings, limitations, contradictions of others, etc.) and explain the concept or theme studied in the lab-work, and justify doing it. It is not a preface, description, or a summary. At the start it must describe the problem in general terms especially in an ergonomics context. Then it should lead on to specific documentation that has been published on the concept, compare and contrast findings, methods, etc., and lead in a funnel shape to a particular topic examined. It should finish in a single sentence stating the exact concept or hypothesis or problem examined in the lab work. Do not say "The purpose of this laboratory was ..." and do not describe here what was done. Just state at the end, in general terms and briefly, the issue that was examined. Divide it into appropriate sections and subsections (e.g., 1.1, 1.1.1, etc.). Total length is to be one page or 300 words (5 letters = 1 word).

2. Method (8% of Marks)

Divide this into appropriate sections and subsections (e.g., 2.1, 2.1.1, etc.) about participants, apparatus, stimuli, design of experiment, procedure, and so on. It must give enough detail for somebody else to repeat it exactly elsewhere. But only include those things that are relevant to the method of investigating the question at hand, that is, might have a bearing on the results obtained. Specific lengthy details should be given in tables, quoting the appropriate labels. It must have appropriate subheadings and paragraphs. The length of the narrative part is one page or 300 words.

3. Results (30% of Marks)

A narrative type of account must be given to summarise the results obtained, quoting data of interest. Lengthy details must be given in tables, and these must be spread through this section at appropriate points in the text, not in appendices. Figures are treated likewise. The text must be succinct and to the point to summarise findings and highlight points. Merely state findings; do not attempt to explain them or speculate on reasons for them. Point out whether they support or reject expectations; give levels of significance and outcomes of statistical analyses (e.g., "… time was significantly different [$t = 5.46$, $p < 0.01$]"). Refer to relevant tables and figures at appropriate points in the text. Write in flowing sentences, without endless minutiae of such things as null and alternative hypotheses and other minor details. Divide it into appropriate sections and subsections (e.g., 3.1, 3.1.1, etc.) with appropriate subheadings and paragraphs. It should be possible to find out what the experiment showed only by reading the text; the figures and tables merely provide extra detail and clarity. The length of the narrative is about three pages (900 words) over and above the tables and figures.

Please note that presentation of the raw data should only occupy a minor part of this section. The bulk of it should consist of the analyses, their presentation, and what they show (including figures). The initial analysis should be at a general level (main effects and tables, etc.) followed by the detailed analyses. Try to find things in the data to demonstrate analytically why experimental factors did or did not affect the results. Analyse the data to see if anomalies can be explained.

4. Discussion (15% of Marks)

This is not a rehash or summary of the results. Divide it into sections and subsections (4.1, 4.1.1, etc.). Try to explain or speculate on reasons for the results. Consider how they relate to the final sentence of the introduction and to previous work in the area given in the introduction (especially). Identify any relevant factors that may cast doubt on the validity of the data. Explain what improvements in procedure (if any) should be incorporated if doing it over again. Consider what else could (or should) have been done. The length is one page or 300 words.

5. Conclusions (4% of Marks)

Do not give a summary of the work but a series of numbered one-line or two-line statements of what the lab work revealed. Confine yourself to the concept and experiment itself, not to generalisations about ergonomics or things known outside of the experiment. Make sure you use the past tense; the present tense implies universal truths. The length is about half a page or 150 words.

References (4% of Marks)

Here, specify the references specifically quoted in the text. The format used must be that followed in the laboratory exercises. Use initials only, not first names, and do not use "Jnr, Snr", etc. Where there is more than one author for a particular reference, all names must be given in full, that is, do not use "et al." here. It can also include a bibliography of general background material. Length is half a page to a page.

Appendices

For large amounts of data, details of unusual tests, etc., not for tables or figures.

Writing Style, etc. (12% of Marks)

Correct grammar must be used with proper sentences, not a telegraphic or military style. The work is history, so past tense must be used in describing what you and others did. Use the present tense to describe theories about human performance and to refer to figures and tables in the report. Use the third person (i.e., do not use "I" or "we" or "you" but "It is thought that ..." or "Participants found that ...." or "... the author ...", and so on). Avoid repetition, waffling, padding, and rambling discourses. Referencing in the text must be as: "Fitts (1954) said that .....", or "Fitts and Posner (1966) state ....", and so on (see *Ergonomics*). If there are more than two authors use the first one only followed by "et al." Do not use abbreviations such as "Fig." for Figure". If the same author appears twice in a year add a, b, etc., after the year (e.g., Smith [1988a]). Do not just copy out the lab sheet or the notes or material from the papers or their abstracts, but use your own words except where some short quote is especially good. Use a proper series of paragraphs (more than one sentence) and flowing sentences with a clear flow of thought linking from one paragraph to the next. Make it clear and easy to understand, organised, and coherent. Do not give opinions but use data to argue a case.

Understanding (12% of Marks)

There must be a clear grasp of principles, ideas, and implications.

Report Presentation

See the specific details spelled out in 1.3 above under "Report Presentation".

Caution: Does the manuscript comply with the Grading Scheme questions and the Report Presentation form? Have all the references from the text been inserted in the Reference section, in the correct format? Check for typographical, spelling, and grammatical errors. Do table and figure numbers tie up correctly with the text?

1.5 GRADING SCHEME FOR LONG REPORTS

		POINTS
Introduction:	Is the subject introduced relative to the broad field of work?	
	Is the survey of previous work broad enough and critical enough?	
	Is the idea clearly and logically developed in a funnel shape?	
	Is the hypothesis/purpose clearly stated?	15

Method:	Is it complete? Is it specific and reproducible?	8
Results:	Do the figures and tables have the correct format and labelling? Have all the analyses been completed? Are the graphs correctly and clearly drawn? Have all the results been presented? Are they all summarised and correctly stated in the narrative? Are significant items and/or unexpected results pointed out? Have levels of significance been calculated and clearly stated?	30
Discussion:	Are the implications of the results interpreted? Have all the questions been explored? Have expected and unexpected outcomes been examined and explored? Have the results been compared to those in the literature?	15
Conclusions:	Are they warranted by the data? Are they complete and accurate? Was the experiment needed to discover them?	4
References:	Has the correct format been used? Are all the references quoted in the text listed here? Are all the references listed here quoted in the text?	4
Writing:	Is the style correct, not flowery nor telegraphic nor colloquial nor military? Is the past tense used and are the tenses consistently correct? Does it have proper sentences and paragraphs and is it intelligible? Are the grammar and spelling consistently correct? Are trains of thought clear and easy to follow?	12

Understanding: Has the student understood the literature?

Has the student grasped the important points?

Is the subject matter understood adequately?

Does the student understand how these results relate to the literature?

Does the student understand the implications of the 12
work?

NOTE: Copying other people's work and presenting it as your own (i.e., plagiarism) is a very serious offence and violates the Declaration that has be signed and submitted with the report. Doing so will incur serious penalties.

1.6 PROFESSIONAL TYPE REPORTS

This type of report should have the following format:

Front page: At top of page: title of the study.
The top half describes what is being investigated.
The lower half describes "Recommendations".

Second page: The top half describes and is titled "Methodology" and refers reader to the appendices for details.
The bottom half spells out a summary of the "Findings" (i.e., what came out of the assessment) and refers the reader again to the appendices for more detail.

Other pages: Consist of the Appendices, which give the details of all of the methodology, detailed findings, tables of the data collected, calculations, graphs, diagrams, process charts, photographs, etc.

The aim is to be succinct in presenting data collected without a lot of reference to theory and/or scientific literature. For this reason such reports are appropriate when reporting on applications of existing knowledge or methods when investigating or addressing an ergonomics problem, such as in field studies. They are not suitable for work examining a theory or a methodological problem.

Reporting style: The writing style and form of presentation must conform to that already specified for the other forms of report.

1.7 INFORMED CONSENT

To comply with human ethics requirements all participants must be given a detailed explanation of what the work entails (preferably in writing), and all questions must be answered as fully as possible. For ordinary laboratory experiments that will usually be sufficient, provided the protocols and procedures comply with general guidelines. For more detailed investigations (such as for the project or thesis) a more detailed and comprehensive document must be presented to the Human Ethics Committee for approval.

To the Signee:

Before signing this form you must be provided with a written description of the experiment in which you are about to participate. The Department guarantees that, if you do decide to participate, all of your data will be kept confidential. No unauthorised individuals will have access to it and any forms connecting you with specific data will be destroyed after the entire experiment has been completed.

Agreement:

I, ____, have read the description of the experiment in which I am about to participate that was provided by ________________________________and all my queries have been satisfactorily answered. I understand that I may freely decline to participate in any part of it and that I am free to withdraw from the experiment at any time during its conduct.

_______________________ _______________________

(Date) (Signature)

1.8 WRITING UP A THESIS OR PROJECT REPORT

It must consist of the following parts and chapters, in this order, and be constructed as specified, but the Introduction and Method may be split into two or more separate chapters, labelled differently. All chapters must be given a descriptive label or title. For general guidance consult learned journals such as *Ergonomics* and *Human Factors*. Use upper case (bold) for chapter headings, upper and lower case (bold) for other headings, as in this document.

Front Section

Title page: Full title, total number of volumes (if more than one), number of this volume, full names of candidate, ID number, the award for which it is submitted, "University of XXXXXX", "Supervisor" and name (s) of supervisor (s), and the statement "Submitted to the University of XXX, (month), (year)". (*Note:* quotation marks mean this is the text to be used. Do not include the quotation marks in your document. Also, volume does not equal copy.)

Second page: "**DECLARATION**"—Full title, "Supervisor" and name (s) of the supervisor (s), and this statement: "This (project, thesis) is presented in partial fulfilment of the requirements for the degree of — (e.g., Master of XXX in Ergonomics). It is entirely my own work and has not been submitted to any other university or higher education institution, or for any other academic award in this university. Where use has been made of the work of other people it has been fully acknowledged and fully referenced". "Signature" (and signed)

Printed full name of candidate
Day, month, and year

Third Page: **DEDICATION**, if any.

Fourth page: **ACKNOWLEDGEMENTS** to people whose efforts helped in the work undertaken such as the supervisor, participants, funding agency.

Fifth page and following: **CONTENTS**: detailed listing, in sequence, of all chapters, sections, subsections, parts of subsections, with page number where each one starts. It includes each individual appendix and/or the major parts.

Next part: **LIST OF FIGURES** with number, title, page on which it occurs, for each figure.

Next part: **LIST OF TABLES** with number, title, page on which it occurs, for each table.

ABSTRACT

This summarises what was done, how, where, and with what results; it enables a reader to decide if it is relevant to his/her interests. It is written after everything else and is completely self-contained. It should be 300 words maximum, single spaced, and give the author and title as a heading.

1. INTRODUCTION

This is the first chapter, it is more than a preface, and it is not just general waffle about the subject. It sets the scene and gives a broad view of previous work, the context of the present work, and a literature review of relevant subject matter. It must explain the concept or theme studied and the justification for carrying out the work. It is not a description or summary of what was done. It must not just recount what others did or said, but must critique the published material, suggest reasons for different approaches and results, compare and contrast findings, and point out gaps, inconsistencies, doubts, or contradictions.

At the start it introduces and describes the problem studied in general terms in relation to the field, with background descriptive material and the origins of the work. Then it leads on to specific research literature, etc., that has been published on the topic and leads in a funnel shape to the particular issue studied (think of a tree diagram). Break it up into appropriate sections and subsections (e.g., **1.1 Background**, **1.1.1**…, **1.1.2**…, …, **1.2**…) and finish in a paragraph or sentence stating the exact issue examined, but in general terms, not as a summary of what was done.

2. METHOD

There must be enough detail about participants, apparatus, techniques used, analyses, computer techniques/programs, design of experiment, procedure, stimulus material, etc., that somebody else can repeat it exactly. Only include things which are relevant (i.e., might affect the results obtained). Lengthy details should be given in tables quoting Table numbers. Divide it into sections and subsections (e.g., **2.1**

Participants, **2.1.1**…, **2.1.2**…, …, **2.2 Apparatus**, **2.2.1**…, ...). Do not include any results here.

3. RESULTS

A narrative type of account must be given to summarise the results obtained, and the details of their analysis, quoting supporting data and statistics. Lengthy details must be given in tables and these must be spread through this chapter, close to first mention. Figures are treated likewise. All results must be presented, but the treatment must be succinct and to the point and merely state findings. Summarise data in tables; use figures to illustrate the "pictures" given by the data. Presenting the raw data should be a minor part of it.

Analysis of the data should be at a broad level first (main effects and interactions), with tables, followed by detailed analyses to find out the causes of the findings or failures to reach significance. Do not attempt here to explain them or speculate on reasons for them; those go into the Discussion. Point out features of interest (e.g., whether they support or reject expectations); give outcomes of statistical analyses and levels of significance. Refer to relevant tables and figures at appropriate points in the text. Write in flowing sentences and paragraphs, broken up into appropriate sections and subsections (e.g., **3.1 Analysis**, **3.1.1**…, **3.1.2**…, …, **3.2**…, ...). The narrative should give a good overall or summary picture of the Results without the need to look at the tables or figures. The latter merely add detail if required.

4. DISCUSSION

This is not a rehash or summary of the Results. Here you should explain or speculate on reasons for the Results. How do they relate to the final paragraph of the Introduction and the work of others summarised in the Introduction? Examine and explore expected and unexpected outcomes. Use appropriate sections and subsections (e.g., **4.1 Analysis**, …, **4.1.1**…, **4.1.2**…, …, **4.2**…, …). Do your results differ from those published in the literature? Why? Note any relevant factors that cast doubt on the validity or quality of the data, methods, procedure, participants, assumptions, etc. Why? What changes might have improved the work? Why? What else might you have done? Why? Do not repeat results, just say, e.g., "That factor B had no effect may be due to …".

5. CONCLUSIONS

This is not a summary. Give a series of numbered statements (each of a few lines) of what the work revealed. Ensure that each one makes a separate point. Confine yourself to the work itself, not to generalisations or already known results. They must be supported by the work you carried out and the data in the Results. It must not contain speculation or discussion, or repetition of results, or findings known before, or without doing the investigation. Ensure use of the past tense; present tense implies universal truths. This will probably run to two pages or more.

6. SUGGESTIONS FOR FURTHER WORK

Ideas generated by the work, especially after reviewing what was done, are included here. How can it be extended, improved or generalised? What new topics has the work suggested as being worth investigating? This should be at least as long as the Conclusions, preferably longer, and provides an opportunity to show some creativity.

REFERENCES

Credit must be given for all techniques, layouts, diagrams, information, ideas, graphs, formulae, software, algorithms, etc., used from other people, from other documents, or from vendors. When quoting or paraphrasing someone else's work, that work must be referenced at this point in the text. All references specifically quoted in the text must be listed here, and all authors listed here must be quoted in the text. If an author appears more than once in the same year, add a letter after the year (e.g., Smith [1988a] and Smith [1988b]). In this section itself, the format must be that followed below, giving full details of all authors for each reference. The style is as for these:

Hicks, J.A., 1976, An evaluation of the effect of sign brightness on thc sign-reading behaviour of alcohol-impaired drivers, *Human Factors*, 18, 45–52.

Konz, S., 1989, *Work Design: Industrial Ergonomics* (3rd ed.), Publishing Horizons, Scottsdale, Arizona.

McCormick, E.J. and Sanders, M.S., 1993, *Human Factors in Engineering and Design* (7th ed.), McGraw-Hill, New York.

Sakai, K., Watanabe, A. and Kogi, K., 1993, Educational and intervention strategies for improving a shift system: an experience in a disabled persons facility, *Ergonomics*, 36, 219–225.

Full information must be given on all documents listed so that the reader can obtain a copy if wanted. With multiple authors use alphabetical order of names first and, within these, by years. Author alone precedes author with others. They come before the Appendices. Here full last names must be given for all authors but initials instead of first names; avoid Senior or Junior or 3rd or whatever of that sort. Do not use "et al." here; use it only in the body of the text, if more than two authors. Do not use abbreviations for the names of journals or other documents listed here. Note that terms such as vol., no., pp., page, pages, etc. are omitted; they are deduced merely by the positions of the data.

BIBLIOGRAPHY

A listing of general background material can be added after the above, if desired.

APPENDICES

These come after References and Bibliography and contain details of special tests, maybe actual raw data collected, computer programs, test procedures, standards documents, informed consent forms, instructions to participants, and so on. They do not contain tables of results or figures.

OTHER ISSUES

1. Writing Style

Correct grammar must be used with proper formal sentences, not a telegraphic or colloquial or military style. This work and that of others is history so past tense must be used for what was done, what participants did, what previous researchers did. Use present tense to describe details of theories proposed or past, and to refer to figures, tables, and calculations in this document. It must be in the third person, that is, do not use "I" or "we" or "you" but "It is thought that ..." or "Participants found that ...", and so on. Avoid waffling, padding, and rambling discourses. Do not use abbreviations such as "Fig." for "Figure". Sentences must contain a finite verb, must contain clear trains of thought, and must be easy to follow. The format for references in the text must be as: "Fitts (1954) said that ...", or "Fitts and Posner (1966) state ...", and so on. Alternatively it may be as "… information lost in transmission (Shannon and Weaver, 1949) would be …". If there are more than two authors, give the first name followed by "et al." and the year (e.g., Smith et al. (1966). Note that data are plural and there are no commas before or after the year in references in the text.

2. Structure of the Document

Chapters must be broken up into clear sections, subsections, and parts of subsections, each titled appropriately, with decimal numbering. They must be subdivided into paragraphs of more than one sentence. Leave a blank line between paragraphs. Justify pages both sides.

3. Editing and Proofreading

Ensure that sentences and ideas connect intelligibly and clearly to others in the text, that duplications have been removed, or omissions, spelling, punctuation, and typing errors have been corrected and ensure that the text ties up correctly to figures, tables, appendices, references, page numbers, and sections.

4. Units

All units and their notation must conform to the Systeme Internationale (SI). Masses must be in kilograms (kg), forces and weights in Newtons (N), lengths in metres (m) and millimetres (mm), velocities in metres per second (m/s); times in seconds (s), minutes (min), hours (h), days (day). Do not include any units from other systems and do not mix units (e.g., seconds and minutes). *Note:* the unit used is always given as singular in tables, etc.

5. Caution

Before finalising the document get the supervisor to check out the style, structure, writing, and presentation. Also, have all the requirements listed here been met?

2 Workplace Environment

The subject matter of this group of exercises addresses some of the issues concerning Health. It has always been part and parcel of Ergonomics from the very beginning. But in subsequent years, with the development of specialists in Occupational Hygiene, some or all of this material has also been addressed by these people. Likewise, in some countries, it is seen as the province of Health and Safety practitioners. However, it is an integral part of the material in HETPEP and therefore forms an essential part of the material to be covered by this document.

For all four exercises, most educational institutions are unlikely to have specialist laboratories set aside for studies of this type. Hence, they will in many, if not all, cases form a pseudo field study, in that the data will have to be collected in real workplaces, possibly on campus. However, to avoid interfering with the normal work activities, it may be necessary to conduct the studies in non-work periods such as at night or at the weekend. For this reason these exercises provide a particularly useful halfway house between pure laboratory work and real field studies. To complete the learning process it would be ideal if they were followed later on by one or more real field studies where these types of data are collected by some or all of the class acting as an investigation team where individually, or in small groups, they conduct an integrated study of these aspects of the work system. Even better would be a study that included an evaluation of the cross-section of work activities according to the programme spelled out in exercise 6.4.

Another particular characteristic of these exercises is that they involve the use of specialist instruments to measure purely physical phenomena. An advantage is that there is little difficulty about risks or embarrassment to the members of the class or to test participants. In theory the measurements should be exact, but the process will soon make it clear to the participants that exactly repeatable values are hard to achieve on these types of measurement. That will form another valuable learning part of the work, and accentuate the need for repeated correct and precise procedures and attention to detail. In some cases it will help to familiarise them with appropriate national and international standards.

Students are strongly advised to consult the documents produced by the American Industrial Hygiene Association (2700 Prosperity Ave., Suite 250, Fairfax, VA 22031) and their Journal; see www.aiha.org.

PARTICULAR EQUIPMENT NEEDS

The ventilation experiment requires a fan and extractor assembly. But, as most of these are likely to be in use, it is advisable to construct one especially for the purpose. The fan should be connected to an extractor box or hood with an assortment of intake fittings with details as specified in the ACGIH books (1998, 2003), so that the effects of changing the intake configuration can be seen. Some typical examples are given on the Web site for this book.

Note: These exercises are intended to introduce the student to the steps and processes involved in doing such studies, but do not attempt to fulfil the detailed requirements specified by specialist organisations or those required by law.

2.1 LIGHTING SURVEY

OBJECTIVE

- To learn the types of procedures required to carry out such a survey

APPARATUS

Lux measurement meter (e.g., Eurisem)
Photometer, for example, Tektronix (ensure it has been charged beforehand)
Wooden stand for Lux meter to hold it in the correct position or positions
Tripod for the photometer
Tape measure in metres
Extension power cord

TECHNICAL BACKGROUND

High precision work requires high lighting levels (Konz and Johnson, 2008), but they in turn risk causing glare, either from the work piece or work surface, or from the light source itself. But, if the lighting level is too low, the worker cannot see adequately enough to meet job requirements or to avoid mistakes. A compromise is needed, especially when the lighting costs are considered.

Alternatively, many, if not most, workers would like to work with natural lighting and, possibly, a pleasant view outside. Unfortunately, this can result in distractions with consequent negative effects on quality, productivity, and safety, or may result in direct or indirect glare from sunlight and/or reflections from outside or inside surfaces. The compromise in this case is often to place windows at a height above the line of sight of workers when they are performing their tasks, or to mask the windows in some ways. See Howarth (2005) for suggestions.

In the end it means that there is a need to ensure that there is sufficient light falling on the work surface (illuminance), and that there is not too much light (or too little) entering the workers' eyes (luminance). These requirements are further complicated by reflections from the work area and various surfaces in the work space such as walls, ceilings, equipment, furniture, floors, clothing, and so on. Hence, two types of light measurement are needed, and the process must ensure that all sources are examined in any detailed study.

PROCEDURE

1. Prepare the conditions of the environment and the instruments as follows:
   The lighting installation should have had at least 100 hours of operation before starting.
   On the day, the installation should have been lit for at least one hour before measurement.
   The instruments should have been calibrated within the last 12 months.
   The surveyor must be dressed in low reflectance clothing.
   N.B. Leave furniture and equipment in their normal positions as much as possible.
2. Complete the details for Table 2.1.
3. Choose the accuracy required. For the greatest accuracy divide the room/space into a grid of equal sized squares (usually 0.5 m). To ensure that the grid is symmetrical, increase the number of measurement points if necessary. For quicker and less accurate results use squares of 2 m, OR use the averaging procedure given in the IES document (IES Committee, 1963) or those of another similar document.
4. Take illuminance readings of the work area (light falling on the surface, lux) at the centre of each square with the surface of the light-sensitive cell horizontal at these heights:
   0.76 m above the floor for offices (regard a classroom as an office)
   0.85 m above the floor for industrial premises
   N.B. Expose the instrument for at least 1 minute before taking each reading.
   Omit daylight effects—use blinds, or do it at night. (But, to do 5 as well, leave as daylight).
   Record the results on a grid of the work area floor.

TABLE 2.1
Description of the work area

Interior surfaces	Material	Texture	Colour	Reflectance (%)	Condition
Ceiling					
Walls					
Picture rail					
Trim					
Floor					
Shades or blinds					
Work surface					
Equipment					

5. Measure the illuminance (light falling on the surface, lux) of the interior, that is, walls, work surfaces, cupboards, doors, etc., for a typical operator position. Take these measurements at eye level (which can be selected as 1.60 m height) on all walls, ceiling, and floor, or any other surface that covers a large area in the field of view. Record the maximum, minimum, and four other typical values in Table 2.2.
6. For task luminance (light emitted or reflected by the surface, cd/m^2 or nit) take the value at the point of work using the photometer. Ensure that the meter surface is in the plane of the work, that is, horizontal, vertical, or inclined. Take these under actual working conditions, that is, with sun plus lights, etc., in use normally and at the positions in Table 2.3. Mount the meter on a tripod and adjust its horizontal and inclined positions to suit the situation.
7. Make a sketch of the area and show the dimensions, including ceiling height, and show the locations of the windows and the exposure (approximately N, S, E, or W). Indicate entrances and the chief orientation of the occupants, if any. Also show obstructions.
8. Recommend changes and improvements, especially with reference to documents from the professional bodies listed in parenthesis (BS EN 12464-1, CIBSE, IESNA, and AIHA) and explained in the references.

REQUIREMENTS

1. A professional type of report.
2. Record the illuminance values (lux) on a 2 m grid of the floor plan and draw in estimated iso-lux lines.

TABLE 2.2
Illuminance measurements (lux) in a typical operator work area

Work point*	Description of work point	Height above floor (m)	Plane (horizontal, vertical, or inclined)	lux values	
				Total (general + supplementary)	General only
Max					
Min					

*On the lower lines give typical values for the workplace.

TABLE 2.3
Luminance measurements (cd/m²) at some typical locations

Aim of photometer	Location					
	A	B	C	D	E	F
Luminaire at 45 deg above eye level						
Luminaire at 30 deg above eye level						
Luminaire at 15 deg above eye level						
Ceiling, above luminaire						
Ceiling, between luminaires						
Upper wall or ceiling next to luminaire						
Upper wall between two luminaires						
Wall at eye level						
Picture rail						
Floor						
Shades and blinds						
Windows						
Windows						
Task						
Immediate surroundings of task						
Peripheral surroundings of task						
Highest brightness in field of view						

Source: These tables have been adapted from IES Committee, 1963, How to make a lighting survey, *Illuminating Engineering*, February, 87–100, and are used with permission.

REFERENCES

AIHA, *Journal of the American Industrial Hygiene Association.*

BS EN 12464-1, 2002, Light and lighting: Lighting of work places, indoor work places, British Standards Institution, London.

CIBSE, 1991, The Visual Environment in Lecture, Teaching, and Conference Rooms, Chartered Institution of Building Services Engineers, London.

CIBSE, 2002, Code for Lighting, Chartered Institution of Building Services Engineers, London; but also see http://www.cibse.org/.

Howarth, P.A., 2005, Assessment of the visual environment, In *Evaluation of Human Work* (3rd ed.), Wilson, J.R. and Corlett, E.N. (Eds.), CRC Press, Boca Raton, Florida.

IES Committee, 1963, How to make a lighting survey, *Illuminating Engineering*, February, 87–100.

Illumination Engineering Society of North America (IESNA), 120 Wall St, Floor 17, New York, NY 10005 and www.iesna.org which has a big selection of useful materials.

Konz, S. and Johnson, S.L., 2008, *Work Design: Occupational Ergonomics* (6th ed.), Holcomb Hathaway, Scottsdale, Arizona.

2.2 NOISE MEASUREMENT

OBJECTIVES

- To measure occupational noise in two situations
- To see how bad they are
- To produce a noise map
- To get an estimate of hearing damage
- To derive countermeasures

APPARATUS

Sound level meter + tripod
Octave band sound level meter with frequency analyser
Sound measuring anechoic chamber (preferably, otherwise a quiet room)
Powered devices such as food mixer, hammer action drill + steel rod and steel plate, industrial vacuum cleaner, portable grinder
Floor plan of the workshop area
Charged up batteries
Two measuring tapes (or rods of 40 cm and 1.60 m), one for each experiment
Copies of manuals for sound level meters

TECHNICAL BACKGROUND

It is well known that noise can cause deafness, and obviously, Ergonomists must find ways to combat it and reduce such risks. What is less well appreciated is the fact that noise is a stimulant, despite the overwhelming evidence all about us from the common liking for loud music. This has other effects similar to those of stress due to the "fight or flight" reaction to it. This reaction is part of our basic animal reaction to any threat or perceived threat, and the body increases its production of adrenalin to provide the ability to run at maximum speed or stand and fight the threat. In other words, even if the noise level is low enough not to cause early hearing loss, there may be adverse effects due to it raising the levels of adrenalin. Therefore, ideally, everyone should live and work in quiet and peaceful surroundings, and Ergonomists should try to achieve such conditions in all occupational activities.

The issue breaks down into two parts. One part is the general level of noise in the work environment. The other part is the noise in the worker's immediate vicinity due to using a tool or machine of some kind. These two types of noise sources require two types of investigation. Noise in the general environment can be reduced at source by engineering design measures but can also be combated by such other measures as enclosures, sound absorbers on the walls to dissipate the pressure waves, and Personal Protective Equipment (PPE). The use of tools is more difficult as the tool vibration is transmitted up the bones of the arm, and this is accentuated by the tight grip required to hold the tool. Engineering design measures are the only countermeasures against this, but PPE can reduce or prevent the hearing loss problems. However, the latter may depend on the noise frequency as some PPE fittings are better at attenuating certain frequencies than others. This means that the noise frequency needs to be measured, as well as the actual noise level.

The combination of noise level and frequency (or frequencies) gives rise to the idea of "noisiness" or what might be termed the annoyingness of noise. Attempts to achieve a satisfactory measure of this have been mixed, and more work is probably needed before a generally acceptable measure has been developed. One such unit is the Noy, and it is a convenient one to use here.

However, even if the noise level satisfies the legal requirements, there is still a risk of noise induced hearing loss. The risks differ according to such factors as age, duration of exposure, noise frequency, and opportunities for recovery. Tables have been developed to assess the risk of these combined effects and are formulated in the British Standard 5330: 1976 and ISO documents (see Haslegrave, 2005). Such estimates can be used to compare workplaces and help to decide on actions required, for example, rotate people between noisy and quieter work areas.

PROCEDURE

Experiment 1—Get a scale drawing of the work area or make your own, mark it out in squares (0.5 m or 2 m) and number them. Set all the equipment running. At the centre of each square, take readings of the noise level using the meter set on fast response and at a sensor height of 1.60 m. Take readings in both directions along the short and long axes in the horizontal plane of the area, and then vertically in the up and down directions. The mean of these is taken as the noise level in the middle of that square. Record these in Table 2.4.

N.B. To avoid shielding and reflecting of noise, use a tripod, and stand well away.

Experiment 2—Set up the powered devices in an anechoic chamber (or a quiet and isolated room) on a support. Use the octave band meter with settings as specified in the User Manual. Run each piece of equipment, and measure the dBA values in the horizontal plane at mid height on its centreline, at both ends along the longitudinal axis, on both sides of the transverse axis through the midpoint, and then vertically above the centre. Take all of these at a distance of 40 cm from the outside of the equipment (to represent half an arm's length).

Let the meter settle down before noting the reading. Decide the noisiest direction. Then measure the noise level in the noisiest direction at each speed of each device, using filters, etc., as advised by the user manual. Take readings at each octave band by manually cycling through the range from 8 Hz to 16 kHz and waiting for the scale to settle at its maximum value each time. Where the speed is continuously variable (e.g., the drill), run the devices at three speeds (Slowest, Medium, and Fastest). Use the drill with and without hammer action with the rod clamped in the chuck and pressing the whole thing down onto the metal plate to simulate work operations. Record the values in Table 2.5.

TABLE 2.4
Room noise observations (dBA) at 1.60 m sensor height

Place	North	South	East	West	Up	Down	Mean

N.B. *Identify "Place" by the number of the square on the sketch of the floor.*
(For notation purposes consider the long axis to run north–south even if it does not.)

Table 2.5
Noise level readings on powered devices at mid-octave bands (dBA) (at a distance of 40 cm in the noisiest direction)

Device	Speed	16k	8k	4k	2k	1k	500	250	125	62.5	31.3	16	8
Mixer	Slow												
	Med												
	Fast												
Drill	Slow												
	Med												
	Fast												
Vacuum cleaner													
Other													

REQUIREMENTS

1. A professional type report.
2. Using a scale drawing of the workshop floor, mark the mean dBA figure for each square and draw in iso-noise lines on the drawing at intervals of about 5dB from about 60 to about 90. Group the areas or subareas into logical dBA sets of about the same level (± 2 or 3dB). Record the actual values in an Appendix to your report.
3. Draw up tables of your data for each of the tools at the various speeds, frequencies, etc.
4. Plot a graph of noise level at each mid-octave band separately for each machine, showing the points for each of the speeds, with different symbols for each speed. Space octave bands equally and plot the noise levels only over the range recorded.
5. Plot a single graph showing separate points for noise level against the mid-octave band for each of the tools (use the noisiest where it has more than one speed), identifying the points for each differently. Use equal spacings for the octave bands and plot noise values only over the range recorded.
6. Using the Sanders and McCormick Workbook procedure, calculate the noisiness levels (in noys) and the Perceived Noise Level (PNL, measured in PNdB) for each tool.
7. Estimate the Leq for the workshop area using the mean at each square as though these values represented a sample of readings taken in one placc at 10-second intervals. (This is not entirely unreasonable for a technician

who moves around among students). Use the simplified method provided on pages 432–434 of the 5th Edition of McCormick and Sanders.

8. Using the British Standard (or similar) estimate the expected percentage who will suffer hearing handicap due to the noise. For the workshop, assume that students are 22 years old and exposed to the noise for 4 years for 4 hours/day using the noisiest level there. For the technicians, assume an age of 65 and exposure for 45 years for 8 hours per day using the Leq value. For the tools use the ages and exposures for the technicians only, but with the worst noise level for each tool. Present the results of these calculations in a summary table. Compare and contrast the levels of the tools.
9. Examine what measures you would take to reduce noise problems in the workshop, other than personal protective equipment, for example, see p. 448 and Figure 15.12 in the 5th Edition (and pages 613–618) of the 7th Edition of Sanders and McCormick. Consult the *Journal of the American Industrial Hygiene Association.*
10. Take the noisiest hand tool and add its dBA figure to the overall level at the quietest square in the workshop. What will the new overall level be in the area using Figure 15.3 and page 430 footnote in the 5th Edition? Show your calculations for this.

REFERENCES

AIHA, *Journal of the American Industrial Hygiene Association.*

British Standard 5330:1976, Estimating the risk of hearing handicap due to noise exposure.

Haslegrave, C.M., 2005, Auditory environment and noise assessment, In *Evaluation of Human Work* (3rd ed.), Wilson, J.R. and Corlett, E.N. (Eds.), CRC Press, Boca Raton, Florida.

McCormick, E.J. and Sanders, M.S., 1982, *Human Factors in Engineering and Design* (5th Ed.), McGraw-Hill, New York, Chapter 15, especially pp. 428–434.

Sanders, M.S. and McCormick, E.J., 1982, *Workbook for Human Factors in Engineering and Design* (5th Ed.), Kendall/Hunt, Dubuque, Iowa.

Sanders, M.S. and McCormick, E.J., 1992, *Human Factors in Engineering and Design* (7th ed.), McGraw-Hill, New York, Chapter 18.

2.3 OFFICE THERMAL COMFORT

OBJECTIVES

- To gain experience in carrying out such a study
- To see how to do the calculations from it
- To learn how to interpret the results
- To compare Fanger measures to others

APPARATUS

Instrument to measure wet bulb, dry bulb, and globe temperatures, for example, SCANTEC WIBGeT

Tripod for instrument

Measuring tape
Botsball thermometer
Sling psychrometer (e.g., Casella)
Hot-wire anemometer (e.g., VELOCICALC)
Distilled water for wet bulb

TECHNICAL BACKGROUND

Thermal Comfort has long been a bone of contention between different people in the same work environment and can lead to serious conflicts between them. Part of the reason is that it depends on combinations of a number of variables that seem to have different effects on different people. The important variables are dry bulb temperature, wet bulb temperature as a measure of humidity, globe temperature as a measure of radiant heat, and air velocity. Fanger (1970) looked at this by combining a ventilation engineering approach with subjective assessments of various climatic combinations from a large number of people. These are the basis for BS EN 7730 2005.

From his data Fanger established estimates of the percentages of people that would be dissatisfied over a big range of these combinations. One of the interesting features of his approach is that it accepts that no matter what combination is used there will always be a small percentage of people who are dissatisfied. The trick in all of this is to reduce this figure as far as possible by adjustment of the atmospheric conditions. Fanger's work provides the mechanism to decide what variables to change and by how much, by using the tables given in his book or by buying a special instrument manufactured by Bruel & Kjaer or BS EN 7730 2005.

Other researchers have tried to do the same, and therefore it is useful to see how some of them compare with his approach. In this exercise, the two looked at are the Botsball temperature and the Wet Bulb Globe Temperature (WBGT). The WBGT is calculated by a formula that combines the measurements of wet bulb, dry bulb, and globe temperatures with the latter using a 150 mm diameter black globe. The Botsball was developed by Botsford using a 50 mm diameter globe covered with black plastic foam that is kept wet by a small reservoir. He has claimed that this is a more valid and accurate way to measure the thermal properties involved. From this exercise, students may obtain their own experience of these differences and compare them with Fanger's method.

Similarly, it can be argued that the sling psychrometer gives a more representative value than an instrument that has no dynamic element. This is obviously questionable because room air normally moves at a rather low velocity and, to some extent, the purpose of using this instrument is more in order to be able to relate results to those found or recommended in the past. The hot-wire anemometer provides the means of measuring the actual air velocity in the room so that calculations can be made for the actual air velocity rather than that produced when whirling the psychrometer.

All the results can be compared with the figures obtained from the Heat Stress Index (HSI), either in its formula form or from the nomogram version. This was one of the earliest tools developed for judging such conditions and has been superseded by more recently developed tools. It was really developed for heavy workloads in more extreme environments, so is unlikely to be suitable for many normal work environments, and part of the purpose is to look at that aspect.

PROCEDURE

1. Divide the room space into rectangles/squares, usually 2.0 m squares. If the people are sedentary, take the measurements at 0.6 m above the floor level, for people standing take them at a height of 1.0 m. For more detailed studies use heights of 0.2, 0.6, and 1.0 m for sedentary people, and use 0.3, 1.0, and 1.7 m for people standing.
2. Measure the air temperature (t_a), mean radiant temperature (t_{mrt}), natural wet bulb temperature (t_{nwb}), and WBGT and measure air velocity (m/s) at the centre of each square. Allow 5 minutes for the instrument to reach a steady state (especially T_g). Work quickly.
3. Because vapour pressure (or relative humidity) is the same all over a room, calculate it at just one point using a psychrometric chart. To avoid errors due to changes over time, take values in a short time.
4. Establish the activity level and clo-value (e.g., from Tables 1 and 2 of Fanger). If activity levels are high, these increase the relative air velocity, so adjust all velocities for this.
5. Determine the Predicted Mean Vote (PMV) from Table 13 of Fanger.
6. If mean radiant temperature deviates from air temperature, correct the table value by using Fanger's Figure 24.
7. If the humidity deviates to any great extent from 50%, correct the table value by using Fanger's Figure 25.
8. If measurements are taken at three heights, the mean PMV is found from the PMV value obtained at each.
9. For each location determine the Predicted Percentage of Dissatisfied (PPD) value from Figure 27 or Table 15 of Fanger and get the mean for the whole space. The PPD is his suggested "figure of merit".
10. Calculate the PMV for the whole occupied zone as the average of the PMV values for each of the individual locations.
11. Get the sling psychrometer and Botsball temperatures.
12. If the overall PMV is higher (or lower) than zero, it means that the temperature level all over should be lowered (or raised), and the amount of change is found from using Fanger's Figure 23.
13. The mean vote for the whole occupied zone found in 10 or 11 is subtracted from the mean votes determined at each of the locations, and the corresponding PPD values are found from Figure 27 of Fanger. The mean values found give the Lowest Possible Percentage of Dissatisfied (LPPD) in the actual room. The difference between LPPD and 5% is the suggested "figure of merit" for the thermal non-uniformity of the room.
14. If LPPD is too high, the reasons must be examined in more detail to discover possibilities for design improvements. To help this, iso-PMV curves should be drawn using the values from 5 in this list. These give a general

view to help suggest where something is wrong and where design changes are desirable.

15. Get the participants to rate the thermal comfort of the room on the subjective rating scale and compare the results with the data derived from the experiment to get Fanger's estimated PMV values.

REQUIREMENTS

1. A professional type report.
2. Carry out all the calculations in the procedure list and draw iso-PMV lines.
3. Compare Fanger values with WBGT, HSI, Botsball, and sling psychrometer values and the limits suggested in Kroemer (2008).
4. Give your recommendations about the room/space.
5. Comment on the procedures as well as your findings.

REFERENCES

AIHA, *Journal of the American Industrial Hygiene Association.*

BS EN 7730, 2005, Ergonomics of the thermal environment. Analytical determination and interpretation of thermal comfort using calculation of the PMV and PPD indices and local thermal comfort criteria.

Fanger, P.O., 1970, *Thermal Comfort*, McGraw-Hill, New York.

Kroemer, K.H.E., 2008, *Fitting the Task to the Human* (6th ed.), Taylor and Francis, London.

2.4 VENTILATION

OBJECTIVES

- To gain familiarity with measurements of air flow
- To measure air flow at the intake to a hood and upstream of it
- To compare flow patterns for different intake designs
- To compare measured values with those claimed in publications

APPARATUS

Extraction hoods with a specially marked out work surface
Hot-wire anemometers such as Velocicalc, and batteries for each
Aneroid barometer
Whirling psychrometer (e.g., Casella)
Height sticks marked for 0.25 W, 0.5 W, 0.75 W, and W below the hood where W = width of hood
Smoke generator (e.g., Drager Air Flow Tester)

TECHNICAL BACKGROUND

Ventilation in this situation is concerned with extracting harmful vapours and small particle matter (such as dust) from the air close to a worker. Mechanical engineering textbooks show the velocity profiles in the area of the intake to a ventilation hood and the pattern of airflow. These patterns and flows change depending on the shape of the hood. The aim is for the students to see for themselves how the velocity profile changes as the air moves closer to the hood, and how it is different at different points across the intake. One particularly important point is to see how low the velocity is at one diameter or width ahead of the intake face.

Preferably this work should be carried out with a rig that has two or three different shapes and/or configurations of the intake hood. In particular, it is desirable to have an intake situated opposite the operator to demonstrate the flow of air horizontally over the surface of the work area and to contrast this with an overhead extractor hood, as used at times.

As the extractor system consists of ducting and a fan or fans behind the hood, it is desirable to conduct some simple experiments on flow in ducts and on fan performance. The intention of this additional feature is not to provide mechanical engineering skills but to familiarise the Ergonomist with some of the problems and losses in the system in order to provide an introduction to the concerns and language of the mechanical engineer, or building services engineer, who will design such a system. It could even provide the basic skills to specify the fan required for a simple system.

It is assumed that this work is done as part of the study of Occupational Hygiene where students will be introduced to toxic substances and the question of capture velocities for removing noxious fumes, dust, or other particles from the work area. Hence, it may also be supplemented by an experiment on blower systems to provide air to dilute the concentrations of toxic substances and to look at the flow patterns and mixing patterns of the gases.

PROCEDURE

1. Set the fan to run at maximum speed.
2. Use the smoke generator to show visually the air flow directions and changes in velocity.
3. Measure and note the maximum velocity at the hood entrance on its central axis.
4. Measure air velocities in the horizontal plane of the entrance to the hood—at five equally spaced positions across it, and at five equally spaced lines from front to back, that is, Far Left (FL), Half Left (HL), Centre (C), Half Right (HR), and Far Right (FR) at equal fore-and-aft spaces of Full Front (FF), Half Front (HF), Centre (C), Half Back (HB), and Full Back (FB).

   N.B. Make sure that the shields for the hot-wire protector do not obstruct the air flow. Use the "Average" setting and take five readings each time and record them in Table 2.6.
5. Repeat the foregoing procedure but in horizontal planes below the hood of 0.25 W, 0.50 W, 0.75 W, and W using the height stick to get the height setting. Record the data in Tables 2.7–2.10.

TABLE 2.6
Velocities (m/s) in the horizontal plane of the hood entrance

Left to right position	Full forward (FF)	Half forward (HF)	Centre position (C)	Half back (HB)	Full back (FB)
Full left (FL)					
Half left (HL)					
Centreline (C)					
Half right (HR)					
Full right (FR)					

TABLE 2.7
Velocities (m/s) in the horizontal plane at 0.25*width below the hood

Left to right position	Full forward (FF)	Half forward (HF)	Centre position (C)	Half back (HB)	Full back (FB)
Full left (FL)					
Half left (HL)					
Centreline (C)					
Half right (HR)					
Full right (FR)					

TABLE 2.8
Velocities (m/s) in the horizontal plane at 0.5*width below the hood

Left to right position	Full forward (FF)	Half forward (HF)	Centre position (C)	Half back (HB)	Full back (FB)
Full left (FL)					
Half left (HL)					
Centreline (C)					
Half right (HR)					
Full right (FR)					

REQUIREMENTS

1. A short laboratory report.
2. Draw up tables of the velocities. Express velocities below the hood as a percentage of the maximum velocity at the centre of the hood entrance. Use the formula to correct from the "Standard Velocity" given by the anemometer and the "Actual Velocity" at room conditions.

TABLE 2.9
Velocities (m/s) in the horizontal plane at 0.75*width below the hood

Left to right position	Full forward (FF)	Half forward (HF)	Centre position (C)	Half back (HB)	Full back (FB)
Full left (FL)					
Half left (HL)					
Centreline (C)					
Half right (HR)					
Full right (FR)					

TABLE 2.10
Velocities (m/s) in the horizontal plane at 1.00*width below the hood

Left to right position	Full forward (FF)	Half forward (HF)	Centre position (C)	Half back (HB)	Full back (FB)
Full left (FL)					
Half left (HL)					
Centreline (C)					
Half right (HR)					
Full right (FR)					

3. Plot the percentage velocities at the five heights through each vertical plane towards the wall (i.e., on FL, HL, C, HR, and FR). Then draw in lines of approximate constant percentage.
4. Compare the results obtained with those one is supposed to get, for example, is there an inverse square loss of velocity as the distance from the edge of the intake increases? How do they compare to the values in textbooks on Exhaust Openings such as ACGIH, or documents of ASHRAE and/or CIBSE? Check the *Journal of the AIHA*.

REFERENCES

ACGIH, 1998, *Industrial Ventilation: A Manual of Recommended Practice—Metric Units*, American Conference of Governmental Industrial Hygienists, 1330 Kemper Meadow Drive, Cincinnati, Ohio 45240. www.acgih.org

ACGIH, 2004, *Industrial Ventilation: A Manual of Recommended Practice* (25th ed.), American Conference of Governmental Industrial Hygienists, 1330 Kemper Meadow Drive, Cincinnati, Ohio 45240. www.acgih.org

AIHA, *Journal of the American Industrial Hygiene Association.*

ASHRAE (American Society of Heating Refrigeration and Air-conditioning Engineers) *Handbook* and other documents. see www.ashrae.org

CIBSE (Chartered Institution of Building Services Engineers) documents. www.cibse.org.uk

3 Work Analysis

One of the progenitors of ergonomics was Industrial Engineering, particularly the work of Taylor (2005). He made the first major attempts to assess human performance in a scientific manner after the work of Adam Smith (1776) who portrayed the benefits of the division of labour when making pins. To some ergonomists it is anathema to mention any such connection as it conjures up images of "Taylorism" or "Fordism", but the historical connection cannot be gainsaid. Taylor's book describes a series of case studies involving time study, method study, physiology, metal cutting, selection and training, and process planning; almost all the contents are highly relevant to ergonomics.

In more recent years this subject has sometimes been called Work Study or Work Design or even Work Science. The content is somewhat mechanistic, which Taylor warned against, and he recommended a broader, holistic approach with a gradual and sympathetic introduction for trainees. Such detailed study of work activities is necessary and beneficial to productivity in almost all cases. As Taylor puts it, the object is to "secure the maximum prosperity for the employer, coupled with the maximum prosperity for the employee". However, Taylor's approach tended towards micromanaging and paid insufficient attention to the wider issues of workers' activities, personal needs, autonomy, and the ability to contribute to workplace improvement. Of course, ergonomists aim to build in the larger job aspects that make wider use of the worker's abilities and interests, and allow for individual differences.

In particular, ergonomists disagree with the "One Best Way" approach, which developed from a mechanistic model where the human is viewed as some kind of machine where "fuel consumption" has to be minimised by minimising movement and effort. Undoubtedly, there is one best way, for one instant, for one individual, but individuals are different physically and mentally, and workers need to be able to vary their postures and muscle usages over time and to vary their mental activities. During a muscle contraction, blood flow is reduced and can be occluded. If an exertion is maintained for some time (so called "static load"), there is a reduction in freshly oxygenated blood and a build-up of metabolites that will lead to discomfort and eventually to possible injury. These mean there is a requirement for regular body movements and frequent changes of posture, for example, to avoid sitting in one position for any length of time. Also, the layout of workplaces may result in excessive angles at the body joints, which can result in bodily damage, commonly known as Repetitive Strain Injuries (RSI), Cumulative Trauma Disorders (CTD), Upper Limb Disorders (ULD), and various other terms.

From the mechanistic model, people such as Barnes (1980) developed their "Laws of Motion Economy" and these were refined and extended by Corlett (1978, Appendix III) to incorporate ergonomics principles and to establish the order of priority among them. In particular, Corlett expressed the principle that each workplace should incorporate a number of equally good designs that can be used by any individual without having to make major changes to it and without having to adopt harmful postures. Even so, ergonomists accept that many of the purely physical aspects, such as workplace layout, often have an approximately optimal solution, provided adjustability is built in to accommodate different workers. Furthermore, such minutiae must be combined into job designs that include other less mechanistic tasks such as clerical work and work planning.

This chapter addresses some aspects of HETPEP area C.5, Work Analysis, and some part of area F, Professional Issues, in regard to the costs and benefits accruing from ergonomics activities. Hendrick (1996) has shown that large increases in productivity and quality can be achieved by ergonomics interventions, and it behoves ergonomists to be able to demonstrate the productivity gains due to their improvements to the design of the work. The days of justifying ergonomics on moral grounds have long gone, if they ever existed, and ergonomists must learn to speak the language of management. These exercises provide one part of that process, and the mechanics of calculating the cost-benefits are well spelled out by "Cost benefit studies that support tackling musculoskeletal disorders", produced for the Health and Safety Executive in the UK (www.hse.gov.uk/research/rrpdf/rr491.pdf).

The aim of these exercises is to pose deliberately some fairly obvious deficiencies in the way the work area is designed and to suggest some fairly easy ways to implement improvements (e.g., provide a basic assembly jig) but to leave scope for the students to use their imagination to devise further improvements. It is also intended that there should be some element of "experiential learning" in that the students feel for themselves the avoidable discomfort, effort, and inconvenience involved in performing tasks that have been designed in a less than optimal fashion.

The first exercise, Design versus Speed, provides a useful introduction to some of the ideas of design of experiments as well as some ideas on how to extract meaning from data by arranging it in a suitable fashion. Our experience is that it provides a good foundation for subsequent experimental work of this type, prior to having covered the more specialised topics in lectures. To provide a broad introduction prior to the more mechanistic tools, the second exercise exposes the students to a general approach to task analysis. The later exercises relate to some of the simple, more mechanistic methods of examining and predicting human performance on repetitive manual tasks.

It is assumed that each laboratory group consists of three students, and in the later tasks each group assembles a different product. Each student has a turn at assembling 20 sets of components, plus a turn at timing the activity, and a turn at being the recorder for the study. It acquaints them with some aspects of data collection on people's work activities. Over the course of several exercises, they become quite expert in performing their particular set of tasks, but even so the exercises can be used to demonstrate learning curves.

REFERENCES

Barnes, R.M., 1980, *Motion and Time Study* (7th ed.), Wiley, New York.

Corlett, E.N., 1978, The human body at work: new principles for designing workspaces and methods, *Management Services*, May, 20–25, 52, and 53.

Hendrick, H.W., 1996, Good ergonomics is good economics, Presidential Address, Human Factors and Ergonomics Society, Santa Monica, California.

Smith, Adam, 1776, *The Wealth of Nations*, Published by Pelican Books, Harmondsworth, Middlesex, England in 1970 (first two pages of Book 1).

Taylor, F.W., 2005, *The Principles of Scientific Management*, 1st World Library, www.1stworldlibrary.org

PARTICULAR EQUIPMENT NEEDS

For all of the later exercises, sets of light assembly tasks are needed. The aim is to use cheap and readily obtainable assemblies consisting of at least five distinct components to make sets of 20 examples of each assembly. Suggestions for assemblies and their component parts are as follows:

Rope clamp—U bolt, yoke, flat washer (2), spring washer (2), nut (2)
TV aerial connection—body, insert, claw, length of cable (15 cm), cap
Metal plates (50 mm square with a centre hole)—plates (2), bolt, flat washer, spring washer, nut (see Web site for drawing)
Car exhaust pipe clamp—U bolt, yoke, flat washer (2), star washer (2), nut (2)
Hamburger—bun, steak, cheese slice, onion slice, pickle (all made from coloured plastics)
Stopwatch clamp—screw, base yoke, cover yoke, washer, nut
Circuit assembly—circuit board, transistor, resistor, coil, capacitor
Three-pin plug—base, earth pin, neutral pin, live pin, fuse, cover screw
Gate valve—body, pipe olive (2), pipe collar (2), gasket, valve assembly
Switchbox—base, switch, gasket, fibre washer, steel washer, retainer
Envelope stuffing—envelope for A4 sheets, four A5 sheets (each of a different colour)
Pipe connector for water or gas—male, female, collar, seal, nipple

Appropriate bins and other containers are supplied, such as would be used in a normal workplace. One example would be Maxi bins for the electronics industry, which come in various sizes to suit different components.

For each assembly a simple assembly fixture is supplied, leaving room for improvement.

Barnes Pegboard Task (one per group)

Dimensions for these are specified in Barnes (1980) but have been adapted to metric dimensions with the permission of publisher John Wiley and Sons, New York, and given on the Web site for this text. Each group requires a pair of pegboards, 30 pegs, an orientation board, and a box to contain the pins when jumbled together.

Card Dealing Boards (one per group)

Dimensions are given on the Web site, as specified by Barnes but converted to metric units with permission from John Wiley and Sons, New York.

3.1 DESIGN VERSUS SPEED

OBJECTIVES

- To see whether or not working faster gives a greater output improvement than can be achieved by better work design
- To demonstrate some aspects of the needs of good experimental design
- To show how meaning can be extracted from data by arranging it in a suitable manner

APPARATUS

Barnes pegboards (two arranged to form a 5 × 6 array of holes), see Barnes (1980)
Pegs (set of 30, see earlier text)
Orientation board (see earlier text)
Storage box for a set of 30 jumbled pegs
Stopwatch (preferably in centiminutes)

TECHNICAL BACKGROUND

There is often a feeling that workers have to work harder if their work activities are better planned with less slack time or that higher productivity is achieved by making workers work harder. This experiment attempts to compare a normal work rate with a bad design of work to a faster pace of work with the same bad design of work, and both of these are compared to a good design of work at a normal work pace. The principle is one of "Clever work not hard work".

At the same time the students are introduced to some basics of Design of Experiments ideas, in this case to counteract the order effect and fatigue. In doing a series of experiments, the participant is likely to get tired and so take longer to perform a task if it comes later in the order than if it were earlier. On the other hand, the participant will grow in skill the later the same task is performed in the order due to gaining skill. These two changes in performance tend to counteract each other but in an unknown manner. If the orders of tasks can be different for different participants, the changes can be configured to avoid these effects. Hence, it is much better to counter-balance the orders than to choose orders that are randomised, as the latter may result in orders that are not balanced for these two effects.

The task in this experiment is an adaptation of one in the workbook for Konz (1990). There are three conditions, and that means that there are 3! possible orders for performing them, that is, 6 possible orders. Also, in order to test statistically for significant differences between all conditions, the experimental error needs to be known, that is, what statisticians more usually call the residual error. To get a measure of it, at least two measurements of each combination must be made. So, if there are three participants in the experiment and each one does the experiment twice, there will be six sets of results. In other words, we can organise that each of the three participants in a group does two of the possible orders, with the result that all six possible orders will be included *and* the residual error will also be measured.

The other feature of this experiment is that it can show how meaning can be extracted from the data by putting the data in the expected increasing or decreasing order, for example, easiest conditions first and most difficult ones last. If the data turn out not to follow the expected order, further investigation will be necessary to see if there are any errors in recording the data, or to see if the assumption of the expected order is wrong. In these ways students will be alerted to some of the fundamental aspects of looking through data, and a simple statistical test can be used to analyse the differences between the conditions to see if these differences are significant rather than due to the residual error.

PROCEDURE

Each student must transfer the pegs to the pegboard in each of three conditions, as follows:

a. **Poor and normal:** With the pegboard hole chamfers on the bottom, work from Left to Right (Right to Left if Left handed) and from Top to Bottom, using the dominant hand only, with the pegs jumbled in the box. Operate at a pace that can be maintained for 8 hours, assuming that pay is by the hour regardless of work done. Insert the blunt end of the pin, taking the pins one at a time from the box. Do not race. Record the elapsed time from the stopwatch.
b. **Poor and fast:** This is the same as (a) but work at a pace that can be maintained for 8 hours, assuming that pay is by the piece.
c. **Clever and normal:** Turn the pegboard over to have the chamfered holes uppermost, have the pins pre-orientated in the board, insert the rounded end of the pin, and work with both hands (one pin per hand) from the centre out and from top to bottom. Position the orientation board running from Left to Right, up against the back edge of the pegboards and centred on its centreline. The pace is the same as in (a). Record the time.

Experimental Design: To try to balance order effects, all six possible orders of the conditions must be performed in the orders presented in Table 3.1. Ensure that the positioning of the storage box, the orientation board, and the pegboard relative to each other is exactly the same for each group.

REQUIREMENTS

1. A short laboratory report.
2. Make a sketch of the workplace, drawn to scale on graph paper, with dimensions.

TABLE 3.1
Times of groups by participants and orders of conditions (centiminute)

Person	Condition by trial		GROUP																							
			A		B		C		D		E		F		G		H		I		J		K		L	
	1st	2nd	1st	2nd	1st	2nd	1st	2nd	1st	2nd	1st	2nd	1st	2nd	1st	2nd	1st	2nd	1st	2nd	1st	2nd	1st	2nd	1st	2nd
1st	a	b																								
	b	a																								
	c	c																								
2nd	b	a																								
	c	c																								
	a	b																								
3rd	c	c																								
	a	b																								
	b	a																								

3. Give a table of results showing the order used by each participant on each trial and the times obtained, as in Table 3.1.
4. Re-write the data from Table 3.1 into the order of Table 3.2. Perform "t" tests on the data to compare the means of the three conditions (a vs. b, a vs. c, b vs. c) in the "Cond. Ave." column of Table 3.2. What do they show?
5. Is there an order effect of being 1st, 2nd, or 3rd (see "Comb Ave." values) in the sequence? Are there any anomalies in the data, such as condition (b) taking longer than (a)? Note: Due to learning and practice effects we expect that times in the "Ave." columns will reduce going from the top to the bottom, and that times in the right-hand "Ave." column will be less on each line than in the corresponding left-hand column
6. Plot a family of curves for the mean time of each Group (one curve per Group) over both Trials against each condition in alphabetical order on one graph to see if there are any differences between the groups. Use different plotting symbols for each Group in the class.

REFERENCES

Barnes, R.M., 1980, *Motion and Time Study* (7th ed.), Wiley, New York (Ch. 21, 654–659 and 537–538).

Konz, S., 1990, *Workbook for Work Design: Industrial Ergonomics* (3rd ed.), Publishing Horizons, Scottsdale, Arizona.

3.2 TASK ANALYSIS

OBJECTIVES

- To obtain a detailed breakdown of the tasks performed in one particular function
- To identify for each task in it: what is done, where it is done, when it is done (say in the sequence), who or what does it, and how it is done
- To determine what improvements should be made to the equipment and procedures

TECHNICAL BACKGROUND

This analysis often comes after a systems analysis followed by a task description stage. The aim is to look at the actions performed, the controls and displays used, the skills required, and the appropriateness and/or difficulties of the way in which the job is designed. The aim is to try to establish in general terms what is being done, how, and with what, so that the various tasks can be viewed from a broad perspective. Then the analysis aims to examine how suited the tasks are to the abilities of people in general and to these people in particular (Shepherd and Stammers, 2005).

Such an examination can reveal poor matches to human abilities and therefore predict poor performance, or a need for particularly skilled operators, or a need for prolonged training, or an unnecessarily high error rate. Changes can then be made in the design, and a final document can be produced that can form the basis for a train-

TABLE 3.2
Times for all groups by conditions, orders, and trials (centiminute)

Exp Con	Pos. in Seq	FIRST TRIAL													SECOND TRIAL													Comb. ave.	Cond. ave.
		A	B	C	D	E	F	G	H	I	J	K	L	Ave.	A	B	C	D	E	F	G	H	I	J	K	L	Ave.		
a	1st																												
	2nd																												
	3rd																												
	Ave.																												
b	1st																												
	2nd																												
	3rd																												
	Ave.																												
c	1st																												
	2nd																												
	3rd																												
	Ave.																												

ing scheme and/or a set of operation sheets. Ideally, it must be spelled out in such a way that any person who knows nothing about the equipment can operate it correctly without reference to any other document. There are two types:

1. **SEQUENTIAL TYPE**: where a series of step-by-step actions is undertaken and the result is listed as shown in Table 3.3, which is an adaptation by Drury from Singleton (1974).
2. **BRANCHING TYPE**: some tasks have to be repeated several times to reach some required level of result before continuing, and/or there may be a branch (or branches) due to different possible outcomes at particular stages. In each case there has to be a condition point to decide where to go next.

The style of presentation here is the same as is used in flow charts, that is, tasks are listed in boxes (one noun + one verb), then branching by a horizontal diamond containing a question relative to the condition, and a repetition or loop by the same method. The stream of tasks should be subdivided into sections with a heading over each (and underlined) to make identification easier. Use numbers to link the parts of the system to parts of a sketch of the machine or person involved. See Figures 6.1 and 6.2 in exercise 6.1 as examples.

N.B. The flowchart must be supplemented by separate sequential charts that list all the details and are cross-referenced to the appropriate boxes of the branching chart.

FUNCTION STUDIED

Choose the setting up or calibrating of a complex piece of apparatus such as for an oxygen analyser. Whatever is used, it should require repetition and branching, and the instructor should talk through the activities carried out in the appropriate sequence as might be done by a skilled and cooperative operator. Students make a detailed record of what is done, where, when, with what, by whom, and how. They should also be alert to spot difficulties and/or errors or fumbles that should be noted carefully.

REQUIREMENTS (SEE DRURY, 1983 FOR EXAMPLES)

1. A long laboratory report.
2. Provide a numbered or labelled sketch of the apparatus.
3. Give an overall branching chart for the apparatus with each sequential part shown by a single box, on an A3 sheet modelled on the example in Figure 6.1 in exercise 6.1. Supplement it with separate A4 sheets for each sequential part using copies of Table 3.4, with the latter cross referenced to the former, and both cross referenced to the sketch where necessary.
4. Analyse the activities required by each hand and foot and the types of Check required each time, for example as percentages.
5. Perform Critical Questioning (Appendix I) on a major step in the task analysis chart. What are the alternatives to these, and what should be done in this task for these things? See Work Design Check-Sheet for guidance (Appendix II) and see the Web tool referenced.

TABLE 3.3
Example of layout and information recorded in a sequential type of task analysis chart

Operation		Control	ACTION*					CHECK*					Control Problems	Display Problems
No.	Purpose	(No. or letter or part where operation performed)	Both Hands	Left Hand	Right Hand	Left Foot	Right Foot	Vision	Hearing	Touch	Kinaesthesis	Smell, Taste (S or T)		
1. 2. 3. 4. 5. 6. 7. 8. etc.	What is the purpose or reason for the action? Conceptual as used for function in Systems Analysis, i.e., one verb + one noun	A, B, C, etc. or 1, 2, 3, etc. or name of the part. Label these on your sketch of the equipment to cross reference them.					Action taken, stating which hand/s or foot performs the action and how it is done.					What tells H.O. that the action has been performed correctly and/or completed? What process, or event, or sight, sound, feel, smell or taste provides this feedback? Give the details here, e.g., hear the click from a switch that has been thrown, or a key pressed, or a door closed, or a visual pattern changed, etc.	Difficult movements, combined actions, lack of smooth response, lack of feel, dedents not clear, etc.	Contrast or image poor, obscured label, too many images, confusing variety, glare or reflections, excess noise, etc.

* *Mark the appropriate column for each operation to show which limb or sense is involved, using a single letter.*

Source: Adapted from Drury, C.G., 1983, Task analysis methods in industry, *Applied Ergonomics*, 14, 19–28, with permission.

Note: H.O. means Human Operator.

Table 3.4
Sheet for recording task details in the sequential type task analysis

Operation		Control	ACTION					CHECK					Control Problems	Display Problems
No.	Purpose	(No. or letter or part where operation performed)	Both Hands	Left Hand	Right Hand	Left Foot	Right Foot	Vision	Hearing	Touch	Kinaesthesis	Smell, Taste (S or T)		

Source: Adapted from Drury, C.G., 1983, Task analysis methods in industry, *Applied Ergonomics*, 14, 19–28, with permission.

6. Present a discussion of the task analysis with regard to such things as a heavy loading of one sense or limb, ambiguities of operation, obscure Check stages, problems with the Controls or Displays, design not suited to person's abilities, limited variety of persons who can do it, difficult task to learn, elevated likelihood of errors, excessive time required, etc.

N.B. Consult Fitts' (1951) List, Shneiderman's Table (Appendix VIII), and Corlett's Principles (1978, Appendix III).

REFERENCES

Corlett, E.N., 1978, The human body at work: new principles for designing workspaces and methods, *Management Services*, May, 20–25, 52, and 53.

Drury, C.G., 1983, Task analysis methods in industry, *Applied Ergonomics*, 14, 19–28.

Fitts, P.M. (Ed.), 1951, *Human Engineering for an Effective Air Navigation and Traffic Control System*, National Research Council, Washington, D.C.

Shepherd, A., and Stammers, R.B., 2005, Task analysis, In *Evaluation of Human Work* (3rd ed.), CRC Press, Boca Raton, Florida, 129–157.

Singleton, W.T., 1974, *Man-Machine Systems*, Penguin Books, Harmondsworth, Middlesex, England.

http://www.ergonomics.ie/mirth.html provides access to aids developed under the EU funded MIRTH (Musculoskeletal Injury Reduction Tool for Health and Safety) project.

3.3 FLOW PROCESS AND TWO-HANDED CHARTING

OBJECTIVES

- To use the flow process chart type of task analysis to describe a "one-handed" solution for the operations of a person in carrying out a light assembly task
- To obtain an improved solution by critical questioning of the contents of the chart
- To redesign and reanalyse the task using a two-handed process chart type of task analysis

APPARATUS

Twenty sets of components for the light assembly tasks

Storage containers (two per component) arranged in line along the back of the bench

Blank flow process chart (Table 3.5)

A3 enlargement of the critical questioning matrix (Appendix I)

Blank two-handed process chart (Table 3.6)

Stopwatch for each group (in centiminutes)

TABLE 3.5
Flow process chart

Sheet No. of	SUMMARY			
Type: Product-Material-Person (ring)	ACTIVITY	PRESENT	PROPOSED	SAVING
Course: Year:	OPERATION O TRANSPORT ⇨			
Assembly Task:	DELAY D INSPECTION □ STORAGE ▽			
Present-Proposed Method (ring)	Distance (m)			
Lab Group:	Time (min)			
Chart by (all names):	Signatures:			

ELEMENT DESCRIPTION	Qty No.	Dist-ance	Time (cmin)	SYMBOL O	⇨	D	□	▽	REMARKS
TOTALS									

Source: Adapted from *Introduction to Work Study,* 4th (revised) ed., 1992, International Labour Organisation, with permission.

TABLE 3.6
Two-handed process chart

Sheet No. of Chart by (all):	SKETCH OF WORKPLACE LAYOUT
Assembly Task:	
Course: Year: Present-Proposed Method (ring one)	
Lab Group:	
Signatures:	

L. H. ACTIONS	O	⇨	D	☐	▽	O	⇨	D	☐	▽	R. H. ACTIONS

	SUMMARY			
	PRESENT		PROPOSED	
ACTIONS	L.H.	R.H.	L.H.	R.H.
Operations O				
Transports ⇨				
Delays D				
Inspections ☐				
Holds ▽				
TOTALS				

Source: Adapted from *Introduction to Work Study*, 4th (revised ed.), 1992, International Labour Organisation, with permission.

TECHNICAL BACKGROUND

The Flow Process Chart was developed many years ago as a means of listing out the activities performed by a worker, with quantities, times, and movement distances, by means of a standard set of five symbols as follows (ASME, 1972):

- **O Operation**: An object is changed intentionally, assembled or dis-assembled to/from another, or prepared for another process, or information is given or received.
- ⇨ **Transport**: An object is moved from place to place, except as part of an operation
- **D Delay**: The next planned step cannot take place immediately.
- □ **Inspect**: An object is examined for identity, quality, or quantity.
- ▽ **Store**: An object is stored under some control.

In particular, totals of each are compiled for comparison with those of one or more alternative designs of the work so that the simplest combination can be chosen. It has been widely used and is a useful way to highlight excessive emphasis on one or two types of activity. It provides a quick and simple standardised method to get a picture of what is involved in a given task.

Better work design makes use of both hands, and such layouts need to be analysed to show the share of work done by each hand. Similarly, such a layout should be a logical development from any design that utilises mostly one hand. For these purposes, two-handed charting is needed, but it merely records the activities, that is, no times or distances. So the two charts are complementary. For that reason this exercise is devised as a step from "one hand" to "two hands" despite the fact that the two-handed design should normally be the first choice in such designs. It is useful for students to be exposed to both and so to see the gain from the one to the other. The starting layout of the bins is deliberately chosen to emphasise the discomfort and extra effort of a bad layout.

This exercise plays an important part in forming a critical way of thinking about the design of work, that is, making full use of both hands, eliminating unnecessary tasks and delays, shortening distances, and simplifying the tasks involved. Such experiences aim to make the student critically aware of deficiencies and avoidable effort whenever work designs are examined, and help to develop a consciousness of the need for high productivity. This exercise lends itself to the use of digital video recordings of tasks either from industry or from laboratory simulations. As such, the video recordings can also be used for self-directed learning and tutorial problems.

PROCEDURE

Devise a simple one-handed assembly procedure, that is, where the work is done largely by the dominant hand, helped by the other hand. Arrange the components in individual bins in a line along the back edge of the bench. Then one student in the group assembles all the components supplied, working at a pace that can be maintained for 8 hours assuming that pay is by the hour, taking one piece at a time from each bin. A second student takes the time to complete each assembly (i.e., cycle time), so getting 20 times, and, if possible, element times or subgroups of element times (say 2 or 3 elements, but such detail may be difficult for beginners). The third student records the order of operations and any events of note, such as bad parts or a difficult insertion task.

Then the students change roles, disassemble the parts back into the bins, and then the second student assembles the set of 20 components. Repeat this once more for the third student. Ensure that the parts are returned to the same bins to achieve repeatability between students, and ensure that all three use the same method and sequence. Reverse the order of containers for left-handed students.

After the third student is finished, devise a new layout with the parts divided equally between two sets of containers in a mirror image sequence, one set for the left hand and one for the right hand. Arrange each set of containers in a single arc centred on a line through the appropriate elbow joint in the sagittal plane. The components are now assembled simultaneously by each hand moving in opposite directions, symmetrically and synchronously. Get each cycle time for each student on the assembly of 20 sets with this new layout.

REQUIREMENTS

1. A short laboratory report.
2. Compile a description of the task on the Flow Process Chart (Table 3.5) with at least 25 steps (see Kanawaty, 1992, for details). Get the mean time for the cycle; add in the transport distances and totals and list all the actions required.
3. Select four elements of the chart from item 2 in this list and examine and critique them using the Critical Questioning technique (Appendix I). Report the result on an A3 enlargement of the sheet, listing the outcomes of the critiques, labelling the elements as 1., 2., 3., and 4. in each cell of the Matrix.
4. Provide a scaled sketch of the job with movement and separation distances of the existing layout. Measure the distance from the lip of each bin to the assembly point. Number the bins on the sketch (for cross referencing in the report) and specify the bin type, for example, MAXI BIN 10.
5. Devise an improved solution using the facilities that are at hand (with/without minor modifications), for two-handed work arranged in two arcs about the elbow joint. For one of the hands, get the new distances. Describe the preferred solution by means of a Two-Handed Process Chart (Table 3.6) for the proposed method.
6. Suggest Further Improvements to this solution that would be possible if more and/or different (and maybe more expensive?) apparatus were to hand. Apply the Work Design Check-Sheet (Appendix II) and the Web tool referenced.
7. Specify the preferred solution for performing this job (from item 5 in this list) by means of a Standard Practice Chart (Table 3.7) to specify what, where, when, with what and how the worker should do it.

REFERENCES

ASME, 1972, ASME Standard: Operation and Flow Process Charts, American Society of Mechanical Engineers, New York. www.asme.org

TABLE 3.7
Standard practice chart

Description of Operation:	
Group & Names:	**Course:**
Workplace Layout:	
Equipment:	
Procedure:	

Kanawaty, G. (Ed.), 1992, *Introduction to Work Study*, 4th (revised) edition, 1992, International Labour Organisation, Geneva.
http://www.ergonomics.ie/mirth.html provides access to aids developed under the EU funded MIRTH (Musculoskeletal Injury Reduction Tool for Health and Safety) project.

3.4 PERFORMANCE RATING

OBJECTIVE

- To see how good the members of the group are at Rating the pace of work of a worker relative to a defined benchmark
- To demonstrate differences between the Rating performances of individuals
- To illustrate the types of error found in performance Rating
- To improve the rating skills of students in the class

APPARATUS

Video player + rating film
Packs of 52 playing cards
Barnes card dealing boards (see Web site for dimensions, etc.)
Stopwatches

TECHNICAL BACKGROUND

In addition to the layout aspect of work design there is also the question of how much time the job should take, which we need to know in order to plan future production capacities and to work out how many operators the company will have to employ. Obviously the time taken depends on the pace at which a person works, and the observer has to judge whether or not the pace of a particular worker is appropriate. The most widely used system is what Barnes (1980) refers to as "performance rating", which assesses "operator speed, pace, or tempo". There is general agreement that walking is an adequate means of demonstrating what such a pace should look like. Both the swinging of the arms and legs, and the forward movement of the rest of the body, can be used effectively to display a particular pace. These have been put into an agreed form of words (Kanawaty, 1992, adapted in Appendix IV), which are demonstrated and tried out in this exercise. It should be noted however that this ILO schema is based on taking the faster walking pace ("standard") as 100 and the slower one as 75 ("normal"), whereas in the United States the slower speed ("normal tempo" in Barnes, or 3 mph) is often defined as 100, and the faster ("incentive") pace as 125. For U.S. usage the terms need to be rearranged.

To train this skill for judging typical jobs in industry it is desirable to have a set of film or video recordings for which the pace has been rated by a number of experts on typical industrial jobs. These are viewed by the students and feedback is given of the judgements of the experts. But these still entail personal judgements, so a more precisely measurable task is needed. For this purpose a pack of 52 playing cards must be dealt onto a board marked out as specified by Barnes (1980) in 0.50 minute as normal in Barnes, or 0.375 minute in ILO. This task has precisely determined movements, distances, and actions, so a precise time can be set. Actual times taken can be compared to this time to give the actual pace at which the task was performed, and that can be compared against the students' estimates.

To gain more insight into the student estimates, their figures are compared with the "actual" pace used, which requires plotting both on axes with the same scale.

Then a positive diagonal line through the origin will represent perfect agreement. Plotting the results will show that some students consistently estimate too high or too low (the plots lie parallel to the diagonal, above or below it, respectively) or they estimate too high at a slow pace and too low at a high pace (the tendency towards the mean), or vice versa, where the plots cut the diagonal near the middle. The experiment highlights these leanings or errors and helps the students to correct what biases they may have.

PROCEDURE

1. Review rating table information (Appendix IV).
2. Perform initial rating exercises on walking: a student walks along the passage outside the Laboratory (say) at a steady pace, and the students estimate the rating and record it. The lab instructor gets the "actual" rating of the pace from the stopwatch time relative to the distance between timing marks (e.g., pillars). It starts with two practice walks followed by 10 test walks at randomly different paces, using two or three volunteers in turn (see Table 3.8).
3. Review the rating film: each student estimates the rating of the pace and records it. The first few tasks shown are used as training, that is, students write down their estimates for each in Table 3.9 and then they are given the "actuals" after each one. On the remainder of the set the "actuals" are only given at the end. Enter these data also in Table 3.9 and then plot Figure 3.1 to view errors and improve rating skill.
4. Card dealing: Start with a few practice deals. One person deals the cards, another estimates the rating of the pace (sitting opposite) and the third takes

TABLE 3.8
Results of walking ratings

Trial #	1	2	3	4	5	6	7	8	9	10
Estimate										
"Actual"										
Difference										

Systematic error = Absolute error =

TABLE 3.9
Results of ratings of video examples

Scene	1	2	3	4	5	6	7	8	9	10	11	12	13	14	15
Est. rating															
"Actual" rating															

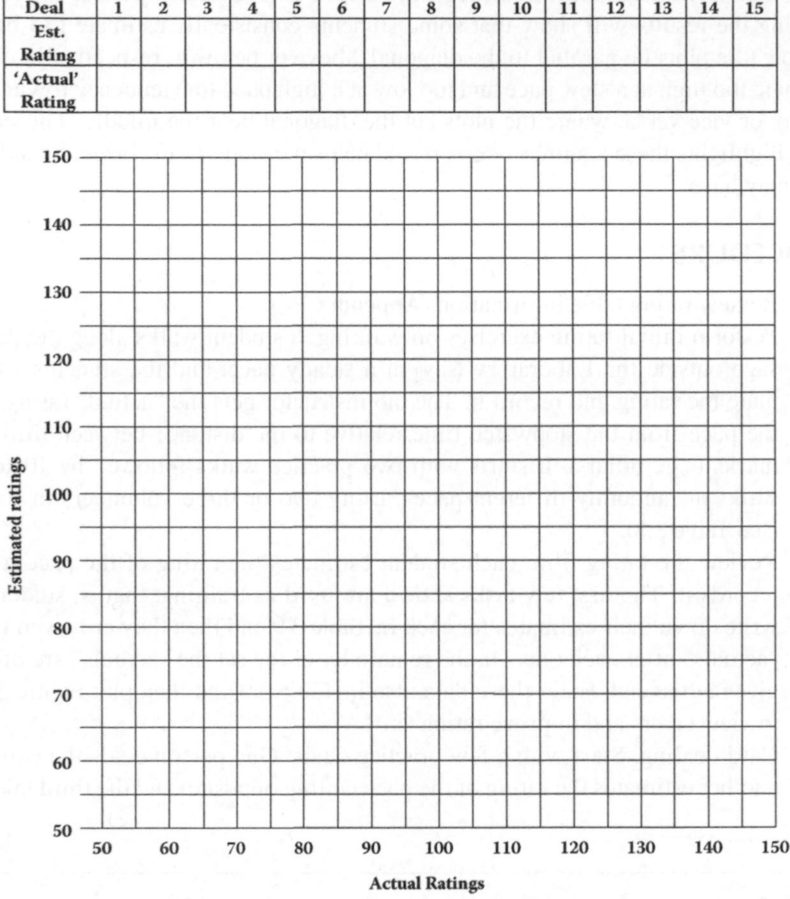

Deal	1	2	3	4	5	6	7	8	9	10	11	12	13	14	15
Est. Rating															
'Actual' Rating															

FIGURE 3.1 Plot of estimated ratings against actual values.

the time (sitting next to the dealer). The deck is held in the left hand (right if left handed), and the top card is positioned with the thumb and index finger of the same hand. The other hand grasps the positioned card, carries it to rectangle 1 and places it on the board, then the next card to rectangle 2 and so on for 3 and 4, then starting again at 1. The cards must all be face down, and each pile must be separate from the others. The dealer must move the dealing hand to each rectangle to ensure that the whole movement is made. The dealer does this for 10 deals (set of 52 cards each time) and does each deal at a steady pace, but makes it different for each deal in a random pattern kept secret. At the end of each deal the timer shows the time on the watch to the dealer and its rating, but not to the rater, and records the rating corresponding to it, NOT the time taken (see Table 3.10 for the conversion). The rater student records their estimate of the rating separately (table in

TABLE 3.10
Conversion of watch readings to ratings for card dealing

Time (cmin)	75	74	72	71	69	68	67	66	65	64
Rating	50	51	52	53	54	55	56	57	58	59
Time (cmin)	63	62	61	60	59	58	57	56	55	54
Rating	60	61	62	63	64	65	66	67	68	69
Time (cmin)	54	53	52	51	51	50	49	49	48	48
Rating	70	71	72	73	74	75	76	77	78	79
Time (cmin)	47	46	46	45	45	44	44	43	43	42
Rating	80	81	82	83	84	85	86	87	88	89
Time (cmin)	42	41	41	40	40	40	39	39	38	38
Rating	90	91	92	93	94	95	96	97	98	99
Time (cmin)	37.5	37	37	36	36	36	35	35	35	34
Rating	100	101	102	103	104	105	106	107	108	109
Time (cmin)	34	34	34	33	33	33	32	32	32	32
Rating	110	111	112	113	114	115	116	117	118	119
Time (cmin)	31	31	31	31	30	30	30	30	29	29
Rating	120	121	122	123	124	125	126	127	128	129
Time (cmin)	29	29	28	28	28	28	28	27	27	27
Rating	130	131	132	133	134	135	136	137	138	139
Time (cmin)	27	27	26	26	26	26	26	26	25	25
Rating	140	141	142	143	144	145	146	147	148	149
Time (cmin)	25	25	25	25	24	24	24	24	24	24
Rating	150	151	152	153	154	155	156	157	158	159

Figure 3.1). The process is repeated twice, turn and turn about so that each student does each part. Plot these in Figure 3.1 with different symbols.

REQUIREMENTS

1. Prepare a short laboratory report and comment on the differences between students.
2. Provide a dimensioned drawing of the dealing board in millimeters.
3. Plot separate graphs for each student in the group of Estimated Rating versus Actual Rating for the card dealing, with separate sets of points for walking, film, and cards— one for each student. Plot a separate least-squares line of best fit for the card dealing set of data. Use 1 cm per 5 intervals of rating, from say 60 to 140. Report the ranges of slope and intercept in the Results (plus other points) and comment on their meaning in the Discussion.
4. Provide a data page with a single table giving the Estimated and Actual Rating figures for each student, with the differences (Estimated – Actual), the systematic error, and the absolute error for each. Also give the regres-

sion parameters for all lines. How does the systematic error compare with C? Why? How do students compare?

REFERENCES

Barnes, R.M., 1980, *Motion and Time Study* (7th ed.), Wiley, New York (Ch. 21, 654–659 and 537–538).

Kanawaty, G., 1992, *Introduction to Work Study* (4th ed.), International Labour Organisation, Geneva (Chapter 22).

N.B. Systematic error = Mean of Differences between Actuals and Estimates
Absolute Error = Difference between Absolute value of largest positive error and Absolute value of largest negative error

For least squares line-of-best-fit: use a software package or the following formulae:

$$E = m.A + C$$

where

E = Estimated Rating on y axis
A = Actual Rating on x axis

$$m = \frac{\sum(A_i \cdot E_i) - (\sum A_i)\cdot(\sum E_i)}{N \cdot \sum A_i^2 - (\sum A_i)^2}$$

and

$$C = (\Sigma Ei - m.\Sigma A_i)/N$$

where

N = number of observations
C = intercept

3.5 OPERATOR LEARNING CURVES

OBJECTIVES

- To examine how performance time decreases with experience
- To estimate the improvement equation parameters
- To find the operator improvement rates
- To see how improvement rates differ between individuals

APPARATUS

Light assembly task
Stopwatch (in centiminutes)
Clipboard
Video machine
Rating video

TECHNICAL BACKGROUND

Improvement in task performance over time was demonstrated by Wright (see Konz and Johnson, 2008) who developed his equation to express it mathematically (see the following text). For simplicity of calculation he plotted the results of time versus cycle number on log–log paper, which gave a straight line. Such plotting has the characteristic that doubling the length along an axis (say in millimetres) doubles the quantity, for example, number of cycles. The percentage reduction in time between two such numbers of cycles was defined by Wright as the "improvement rate" and is the figure used usually for comparison purposes. Barnes (1980) has some nice examples. In this experiment data are collected to examine some of these topics using the light assembly tasks.

By this stage students will have good familiarity with their tasks but not a great deal of skill. So there will be adequate room for improvement without the need to spend any time gaining initial practice on the assembly task. However, their pace will vary a little from cycle to cycle, so it needs to be rated on each cycle to get the Basic Time to get a true measure of the learning effect. Although these latter times will vary, the general trend will normally show an improvement over the time of the exercise despite its brevity.

One of the complexities that can be illustrated here is the difference in improvement rates between different people. Normally these are very clearly visible in the resulting data and cast doubt on a common idea of specifying a global improvement rate for all people on a particular job, that is, that improvement is related to the job rather than the person doing it. The fallacy of a global rate will be demonstrated if the results of one group are combined, and this improvement rate is compared with those of the individual members of the group. It emphasises the importance of individual differences in their abilities and in their patterns of learning manual tasks.

Learning curves can also be useful in showing how differences in work design can have an effect on the rate of learning if the tasks are difficult to learn rather than being easy. Hence, inferior design can result in a lengthy delay before full productivity is reached with attendant extra costs. Alternatively, the extended time may be the result of an inappropriate allocation of functions or of an inappropriate selection of people to do the work. These considerations help further in showing the effects of work design on productivity.

PROCEDURE

1. Review what is meant by cycles and rating (see Kanawaty, 1992, Appendix VI).
2. Refresh the ideas of rating by viewing a rating film. Plot estimated values against actual to sharpen judgement.
3. Set up the assembly task for a largely one-handed operation, that is, where the dominant hand does most of the work and the other just helps (students generally find a two-handed task too confusing at this stage).
4. One member of the group performs the task for 40 assemblies at standard pace (as for 8 hours when paid by the piece). A second student rates each cycle and times each. This must be done standing up opposite the operator

and with the eye, the watch, and the operator's fingers lying on the same straight line. Be sure to do the rating before recording the time. During this activity the third student dismantles the assembled components and puts them back into the bins. Record the results in Table 3.11.

5. Repeat this with each of the other students, turn and turn-about.

REQUIREMENTS

1. A short laboratory report.
2. Present one graph showing the basic time for each cycle for each student, with the points joined by straight lines. Identify separately the points of each student from the others. Suppress the zero, that is, if values run from 15 to 25 have no scale below 15.
3. Provide a table giving the Rating, Observed Time, and Basic Time for each cycle for each student in your group.
4. Provide a table with the mean, standard deviation, and Coefficient Of Variation (COV) for the cycle time of each student for the first 10 cycles, second 10, and third 10. Does variability reduce?
5. Using the first 32 cycles determine the parameters of the improvement curve (a and b) for all three students individually and combined, using simple regression with logs of times and cycles, and give pertinent calculations. Use this to estimate the improvement curve value for each case (say from cycle 10 to cycle 20).
6. Use Wright's equation to predict the time for the 39th cycle. Compare it to the mean of the times for the 38th, 39th, and 40th cycle for each student in the group individually, and all combined.
7. Using log–log paper (or just log values of the data) plot all the points for all students, get a line of best fit by eye, and get the improvement rate for each person in the group doubling from cycle 2 to 4 and from 4 to 8. As a check do the same from cycle 8 to 16 and 16 to 32. Take the mean of these two calculations if the results are similar. Then do it all again taking the combined data of all in the group, and see what results.
8. Present a small table to summarise the improvement data from regression.

REFERENCES

Barnes, R.M., 1980, *Motion and Time Study* (7th ed.), Wiley, New York.

Kanawaty, G. (Ed.), 1992, *Introduction to Work Study* (4th ed.), International Labour Organisation, Geneva.

Konz, S. and Johnson, S.L., 2008, *Work Design: Occupational Ergonomics* (6th ed.), Holcomb Hathaway, Scottsdale, Arizona.

EQUATIONS FOR REGRESSION

Wright's equation: $t = a \cdot c^{b}$

t_i = basic cycle time for cycle i

TABLE 3.11 Data collected for ratings and times (centiminute)

Cycle number	Cycle rating	Cycle time	Remarks
1			
2			
3			
4			
5			
6			
7			
8			
9			
10			
11			
12			
13			
14			
15			
16			
17			
18			
19			
20			
21			
22			
23			
24			
25			
26			
27			
28			
29			
30			
31			
32			
33			
34			
35			
36			
37			
38			
39			
40			

c_i = cycle number for cycle i

so $(\log t_i) = (\log a) + b.(\log c_i)$ gives the equation of a straight line

$$\text{where } b = \frac{n\Sigma(\log c_i)\cdot(\log t_i) - \Sigma(\log c_i)\cdot\Sigma(\log t_i)}{n\Sigma(\log c_i)^2 - \Sigma(\log c_i)^2}$$

and

$$(\log a) = \frac{\Sigma \log t_i - b\Sigma \log c_i}{n}$$

that is, take the log of each cycle time and the log of each cycle number, then take squares, sigmas, etc., or just use software for it.

Remember

Basic Time = (Observed Time) × (Rating/100)

3.6 WORK SAMPLING OR ACTIVITY SAMPLING—SIMULATED

OBJECTIVES

- To use work or activity sampling to estimate four office tasks
- To show how data variability differs across states and observers
- To show how accuracy improves with sample size
- To plan a sequence of observations

APPARATUS

Horizontal time chart of activities of 16 office workers (Table 3.12)
Table of two-digit random numbers from 00 to 99

TECHNICAL BACKGROUND

In many ergonomics investigations one or more groups of people have to be observed, as a study in itself, or as a means to establish which activities form the major part of the work of the group. This often occurs in field studies. From such a study the investigation can be focused where it will be most useful. Also, if some form of stratified sampling is to be undertaken, it is necessary to know how to weight the sampling activity between the various "strata". The problem is how to get these kinds of data.

Industrial engineers developed this technique to meet such a need (Barnes, 1980). The theory is that, by observing a group at random intervals, an accurate estimate can be made of the proportion of time devoted to a task (p). Later this was extended to several tasks. However, randomness can itself result in bias if very extensive numbers of visits are not feasible. To avoid these biases, a better idea is to stratify the sampling so that equal numbers of observations are made on each day

TABLE 3.12

Time chart of 16 people working in an office performing four types of activity

Office employees	Bar charts of actual time at work on the activities by the different employees over one typical work period
1	************************&&&&&&&&&&&&&&^^^^^^^^^^^^^???????????????=********&&&&&&&????????????
2	&&&&&&&&&&&&&^^^^^^^^^^^^^^^******************************???????????????&&&&&&??????????????????
3	????????????????????????????????&&&&???????^^^^^^^^^^^^^^^^^^^^^^***********************&&&&&&&&&&&&&
4	^^^^^^^^^^^^^^^^^^***************************&&&&&&&&&&&&&&&&&????????^^^^^^^^^??????????????????
5	????????????????********************&&&&&&&&&&&&&&&&&&^^^^^^^^^^^^^^^^^^^^&&&&&&&&&&&&&?
6	&&&&&&&&&&&&&&&&???????????????????????????????????*************???????????????????????***************
7	*******************??????????????????^^^^^^^^^^^^^^^^^************&&&&&&&&&&&&&&&??????????????????????
8	?????????????************************^^^^^^^^^^^^^^^^^^^^^&&&&&&&&&&&&??????????**************&&&&
9	**********??????????????????????&&&&&&&&&&&&&&&&&&&^^^^^^^^^^^************&&&&&&&&&&^^^^^^^^
10	^^^^^^^^^^^^^^^^^^^^^^^????????????&&&&&&&&&&&&???????????**************=**********^^^^^^^^^^***********
11	??????????&&&&&&&&&&&&&&??????????*****************************^^^^^^^^^^^^^**************^^^^^^^^^^
12	^^^^^^^^^^????????????***************************&&&&&&&&&&&???????????????&&&&&&&&&*********?????
13	&&&&&&&&&&?????**********^^^^^^^^^^?????????&&&&&&&*********************????????^^^^^^^^^^^^^^^^^^^
14	^^^^^^^^^&&&&&&&&&&&****************************?????????^^^^^^&&&&&&&***?????????????^^^^^????????
15	*********************??????????????&&&&&&&&&&&*****???????^^^^^^^^^^^^^^^^^^^^^^^^^*****?????&&&&&&
16	&&&&&&^^^^^^^^^^^?????????????^^^^^^^^^^^^***********************???????????&&&&&&&&&&&????^^^^^^^^^^

!__________!__________!__________!__________!__________!__________!__________!__________!__________!__________!

% of Work Period 0 10 20 30 40 50 60 70 80 90 100

Activities:

Using the computer	Symbol:	******************
Photo copying		^^^^^^^^^^^^^^^^^^^^
Searching records		?????????????????????
Talking on the phone		&&&&&&&&&&&&

(With acknowledgements to our former colleague, the late Professor Graham Hitchings, who provided the basic idea for this).

of the week and in each quarter of the day or shift. Random intervals are then used within each of these time segments.

However, as always in Statistics, it is necessary to decide on the level of confidence required in the resulting data. To decide on this, it is in turn necessary to decide on the limits of accuracy (L) desired in the resulting estimates. Once these have been chosen, the required number of observations (N) can be determined. In practice, particularly in the days prior to pocket calculators, the level for each of these was set at generally agreed values, that is, 95% confidence and 5% limit of accuracy (or error). Then the equation for the number of observations required (N) was determined (see the following text).

By taking successive samples the students can see how the variability of the estimates of N reduces as the sample size grows. In addition, it is as well to test statistically whether or not the differences between the p values for different activities are significant.

PROCEDURE

All participants do all of the following steps.

1. Get an initial estimate of p: pull 10 two-digit random numbers from the random number table or use a calculator. If using the table, select a spot at random and then read off ten successive different numbers moving along the row or down the column. Alternatively, use a series of seed numbers for the calculator.
2. Treat the % of Work Period line in Table 3.12 as percentages of the working time over the day and then mark off these percentages on the bottom axis. (These correspond to the times for making 10 "tours" of the plant or office). From each point on the bottom line draw up a vertical line to cut the bar chart for each employee.
3. By simple counting, total up the number of occurrences of each state for the set of employees and express these as percentages of the total number of observations (i.e., of 160). If landing on a join, take the LH or RH activity but mark which one so as to avoid double counting. Check that the number of occurrences observed totals up to 160.
4. For each activity, use the calculated value of p with $N = 10$ to find its corresponding limits of accuracy (L) by re-writing the formula as follows:

$$L = (4p(100 - p)/N)^{0.5}$$

5. For the employee activity with the largest value of L calculate the estimated total number of tours needed to give an $L = \pm 5\%$ with 95% confidence (then the estimates of the other activities will be even better) and use

$$N = 4p(100 - p)/25$$

6. Pull 10 more random numbers, mark these on the bottom axis, draw in new vertical lines in a different colour, and count off the frequency of occurrence of each of the activities for these.
7. Using the combined data of all 20 "tours", calculate the new estimate of each p, the new values of L (with $N = 20$), and the new number of "tours" required, for the state with the largest L value.

REQUIREMENTS

1. A short laboratory report.
2. Present a table giving the data obtained by each person in the group, that is, random numbers, p values as percentages, L values, and estimated number of "tours" required, from $N = 10$ and $N = 20$.
3. Treating the two sets of data collected from all three members of the group as 10 tours on each of six successive days, compile a table giving the estimates for p, L, and total required number of "tours" using the cumulative data (i.e., for $N = 10$, $N = 20$, $N = 30$, $N = 40$, $N = 50$, and $N = 60$) to get six sets of values.
4. Plot one graph of L versus cumulative sample size (as in item 3) for each of the four activities of the employees, marking the points differently.
5. Plot a second graph of p versus cumulative sample size (as in item 3) for each of the four activities of the employees, marking the points differently.
6. Do a chi-square test to find out whether or not the difference in percentage between copying and talking on the phone is statistically significant, and whether or not these differ between the first and second "tour". Use just the total number of occurrences of each state in all 60 observations of the office. The "Expected" number in each case is just half the total sets of scores.
7. Work out a set of stratified observations to cover an 8 hour day with a five day week where each tour of observations takes 5 minutes to complete.

REFERENCES

Barnes, R.M., 1980, *Motion and Time Study* (7th ed.), Wiley, New York.

Statistical Test for Differences between Proportions

Use a 2 × 2 Contingency Table and the Chi-square test. Keep the observations as is but as set out in Table 3.13.

TABLE 3.13
Summary of observation data

Result from observation	1st Stream of observations	2nd Stream of observations	Totals of rows
Doing activity	a	b	a + b
Not doing it	c	d	c + d
Column totals	a + c	b + d	n

Because it is a 2 × 2, we improve the approximation by using Yates's Correction as follows:

The expected frequency in cell "a" if the two streams are independent is given by

$(a + c)(a + b)/n$. Compare this to value "a".

Then

	Replace a by (a + 0.5)	if $(a + c)(a + b)/n$	$> a$ to give a′
OR	Replace a by (a − 0.5)	if ditto	< a to give a′

Similarly

	Replace b by (b + 0.5)	if (a + b)(b + d)/n	> b to give b′
OR	Replace b by (b − 0.5)	if ditto	< b to give b′

And

	Replace c by $(c + 0.5)$	if $(a + c)(c + d)/n$	$> c$ to give c′
OR	Replace c by $(c - 0.5)$	if ditto	$< c$ to give c′

And

	Replace d by $(d + 0.5)$	if $(b + d)(c + d)/n$	> d to give d′
OR	Replace d by $(d - 0.5)$	if ditto	$< d$ to give d′

If the new value differs from the original by <0.5, leave the value as is.

N.B. When a value is changed, amend the others to keep the row and column totals the same.

$$\text{Calculated Chi-square} = \frac{n.(a \cdot d - b \cdot c)^2}{(a+b)(a+c)(c+d)(b+d)}$$

Compare to Chi-square table value @ 5% and one degree of freedom, that is, 3.841.

If Chi-square calc < 3.841 we cannot say it is not the same population.
If Chi-square calc > 3.841 we conclude that the populations differ.

(See, for example, Crow, E.L., Davis, F.A., and Maxfield, M.W., 1960, *Statistics Manual*, Dover Press, New York.)

3.7 FUNDAMENTAL HAND MOTIONS

OBJECTIVES

- To use therbligs to describe a light assembly task
- To analyse it in detail from them
- To devise a better method in detail

APPARATUS

Video recorder that can be "stepped" or digital recordings of tasks and viewing software

Video recording of the light assembly task concerned (two-handed)

TECHNICAL BACKGROUND

Gilbreth elaborated the idea that all manual jobs can be broken down into a standard set of fundamental motions common among all types of jobs, and he called them therbligs; these were adapted slightly by Barnes (1980) as shown in Appendix V. To get sufficient detail on these motions in any specific task or subtask it is necessary to film or video the task at a fast speed while it is being done and then play it back frame by frame. The frames are then examined in detail to see which motions are being employed in the task and to discover where difficulties or errors are being caused. These motion details are then used to devise improvements in work design by eliminating subtasks, changing their allocations, changing the design of tools or adding new ones, performing different subtasks, combining subtasks, providing suitable jigs and fixtures, and so on.

The same techniques are now used widely to study the actions of elite sports players, and various very specialised tools have been developed for high-speed study and analysis. Some of these can transfer the details directly to the computer for sophisticated analyses in even finer detail. Alternatively, the data can be used with biomechanics software to extract information on the velocities and accelerations of human limbs and joints, and to determine the forces acting on them or on parts of the spine or other major body parts. They provide a means for studying the dynamics and kinematics of the human body when carrying out typical tasks. Such sophisticated tools give many new possibilities for ergonomics studies (see Gallwey and O'Sullivan, 2005).

Video recordings are required of each light assembly task performed in previous laboratory work, possibly made when previous students were doing them. From previous exercises the students will have personal knowledge and experience of their own task and so will be well equipped to conduct such a detailed analysis.

PROCEDURE

1. Examine the recording several times first, to get a good idea of what it shows.
2. Then break it down into details to get the constituent fundamental hand motions in Gilbreth's therbligs (see Barnes, 1980 and Appendix V for definitions).
3. Ensure that all appropriate motions are included. Assume that occasional mistakes will not occur in practice because an experienced operator will have performed the job thousands of times.
4. Except for this point, analyse the operation as it is done, not as it would be in the ideal state, that is, include time due to fixture problems, parts that do not fit properly, fumbles, etc.

Note: One of the purposes of the exercise is to discover these problems and devise ways to eliminate or obviate them.

REQUIREMENTS

1. A short laboratory report.
2. Compile Actions Breakdown Charts (Table 3.14) to list out the actions performed by each hand, broken down into the constituent therbligs, with the corresponding alphabetic codes and times. Obtain therblig time by calculating the difference between timer readings at the start and finish of each therblig.
3. From the Actions Breakdown Charts draw up Combined Actions Time Charts (Table 3.15) using the alphabetic therblig codes. These depict the Breakdown form results to scale for the durations of actions performed by each hand, show the total cycle time as it accumulates, and show any delays or hesitations or difficulties.
4. Examine the record obtained and critique it by applying the Work Design Check-Sheet (Appendix II) to the therbligs, but only for those that are particularly relevant, and compare against Corlett's Principles (Appendix III) and the Web tool referenced. Present findings in a condensed version of the Work Design Check-Sheet table.
5. Give a separate summary with the total number of occurrences of each therblig and the percentage of time taken up by each for each hand.
6. From the information collected, devise a new task design and present a proposed Combined Actions Time Chart (Table 3.15) to depict it, that is, an estimated version of what the new one should look like. Adapt the data collected to estimate the new times and give the estimated new total cycle time.

REFERENCES

Barnes, R.M.,1980, *Motion and Time Study* (7th ed.), Wiley, New York.

Gallwey, T.J. and O'Sullivan, L.W., 2005, Computer aided ergonomics, In *Evaluation of Human Work* (3rd ed.), Wilson, J.R. and Corlett, E.N. (Eds.), CRC Press, Boca Raton.

http://www.ergonomics.ie/mirth.html provides access to aids developed under the EU funded MIRTH (Musculoskeletal Injury Reduction Tool for Health and Safety) project.

3.8 PREDETERMINED MOTION TIME SYSTEMS (PMTS)

OBJECTIVES

- To apply MTM-1 and MTM-2 Basic Motions to a light assembly task to get time estimates
- To compare these estimates with stopwatch times and with each other.

APPARATUS

Appropriate light assembly task
MTM-1 and MTM-2 tables
Special recording forms

TABLE 3.14
Actions breakdown chart **Assembly Task:**
Student: **Film or Video No. or Letter:** **Sheet ____of____**
Present-Proposed Method (ring one) **Group & Names:**

Left-hand actions performed	Therblig code	Therblig time	Timer readings	Therblig time	Therblig code	Right-hand actions performed

TABLE 3.15
Combined actions time chart

Student: **Film or Video No. or Letter:** **Sheet ____of____**

Assembly Task: **Group & names:**

Present-Proposed Method (ring one)

Left-hand actions performed	Ther. code	Ther. time	Total time	Ther. time	Ther. code	Right-hand actions performed
Timer						Timer

TECHNICAL BACKGROUND

If the fundamental motions have been identified, it should be possible to establish times for them and to use these to estimate the time required to perform a combination of them in a task. This idea was developed by a number of researchers, but the best known implementation is Methods–Time Measurement (MTM), which is marketed by the MTM Association. Their data were collected by analysis of thousands of recordings of a large variety of job tasks, and these have been distilled into their system for time estimation with a system to account for such factors as distance moved, task difficulty, accuracy, object size, force required, load, and so on.

It has a system for classifying the task movements that are called Basic Motions, which are different from therbligs. These are available in the form of tables, and available sources for these are Barnes (1980) for Imperial units, and Kanawaty (1992) for metric units, but a variety of specialist training courses are available at a price. The original fully detailed version is classed as MTM-1, but simpler and less detailed tables are available in MTM-2 and MTM-3 for use where less accuracy is acceptable, perhaps for a short-term task. It is important to note that the rating of pace used in MTM was not the same system as used by the ILO in Kanawaty with the result that the time values must be adjusted accordingly before being used to set time standards there.

In the past, there was some debate about the validity and accuracy of the MTM values (Dudley, 1968), but these have tended to become disregarded over time. They are widely used now in a great many companies with the result that they have now taken on a life of their own, and therefore it is very desirable for ergonomists to know how they are used. The time values used are so-called Time Measurement Units (TMUs) where:

$$1 \text{ TMU} = 0.00001 \text{ hour} = 0.0006 \text{ minute}$$

In the ILO system of rating this is classed as a pace of 83.

PROCEDURE

1. Set up the task for assembly using the same layout and sequence as for the exercise on Operator Learning Curves.
2. Break up the task into its sequence of successive Basic Motions. Then decide the class appropriate to each Basic Motion, measure the curved distances involved, and figure out any special adjustments required, for example, for mass, dynamic factor, simultaneous motion, etc.
3. Read off the times from the MTM-1 tables for these individual motions, then combine them into the elements used for the stopwatch study and get the cycle time. Remember to multiply by 0.83 to get to an ILO rating of 100.
4. Repeat all of the above using MTM-2.

REQUIREMENTS

1. A short laboratory report.
2. Draw up a table or tables (see Table 3.16) giving your detailed breakdown of the MTM-1 motions and the corresponding times. Remember to check for delays when one hand has to wait for the other to complete its activity. Note that rotation during Move is normally ignored.
3. Complete a table giving your detailed breakdown of the MTM-2 motions and their times.
4. Devise a small table to compare both MTM estimates for elements and cycle times with means obtained from the learning curves study, using the times estimated for the end of 2000 cycles. Get a confidence interval from the stopwatch times (t-dist.) and check the probability of getting the MTM-1 and MTM-2 times purely by chance. Use the Standard Deviation of the least variable person in the group over the last 10 cycles of the learning curves study.
5. Discuss the differences between the times, the problems of choosing the correct motion classes, rating in the stopwatch study, subdivision of motions, etc.

REFERENCES

Barnes, R.M., 1980, *Motion and Time Study* (7th ed.), Wiley, New York.

Dudley, N.A., 1968, *Work Measurement: Some Research Studies*, Macmillan, London, St. Martin's Press, New York.

Kanawaty, G., 1992, *Introduction to Work Study* (4th ed.), International Labour Organisation, Geneva.

TABLE 3.16
MTM recording form

Sheet ____of____

Left-hand actions performed	MTM code	TMU value	Cum TMU	TMU value	MTM code	Right-hand actions performed

4 Information Processing

Students sometimes find it difficult to see the relevance of these experiments to their course of study. However, cognisance of the fact that the origin of much of this knowledge is from the fields of experimental and applied psychology and that the testing of fundamental theories and laws is an important part of education in all domains of science and engineering helps justify its importance. Although the application of the equations may not be routine for a practicing ergonomist, knowledge of the theories expands our general understanding of information processing and interaction design, which is a very important part of the discipline.

Usually, this aspect of human performance is difficult to study without sophisticated equipment. However, all the experiments outlined here are performed with simple apparatus and can be conducted in a classroom. Their aim is to show some of the basic characteristics of how people deal with information received through their perception system, and some of the factors that make it easier or more difficult to process.

In particular, this builds on the work done by Shannon and Weaver, who developed the ideas of Norbert Wiener. Humans extract a set of signals from the barrage of information presented and convert the set into a message, after which the human has to decide to do something or leave things as they are. Each signal S_i has a probability p_i of appearing and the amount of information in the signal is $p_i \log_2 p_i$ bits, so that the total information in the message is $\Sigma\, p_i \log_2 p_i$ over the n signals in the message. This idea was developed further by Hick (1952), who found that when he plotted reaction time (RT) against n equiprobable alternative signals, he got a straight line if he used $\log_2(n + 1)$, where the +1 was needed as the person had to decide also whether or not a signal was present.

So, Hick's law has the form

$$\text{Choice RT} = \text{K}\log_2\text{N}$$

where K is a constant and $\text{N} = n + 1$ for n equiprobable alternatives.

When the number of alternatives is not equiprobable, Hick's law takes the more general form:

$$\text{Choice RT} = \text{K}.\ \Sigma\ {}_i p_i \log_2 p_i \text{ where i} = 1, \ldots, n$$

When these data are plotted, K = slope of a straight line = 1/R, where R = the rate of information transmission (bits/second) = a constant.

Because R is a constant, we can conclude that the human processes information at a constant rate regardless of the amount of information or its form, which is very useful when designing and evaluating these types of human interactions.

REFERENCE

Hick, W.E., 1952, On the rate of gain of information, *Quarterly Journal of Experimental Psychology*, 4, 11–26.

PARTICULAR EQUIPMENT NEEDS

Special cards are needed for the exercise in Section 4.1 as they cannot be purchased currently. They should be 50 mm square with a black disc centred on the cards in sets of 20. Due to the size effect, there should be sets of "standard" discs at diameters of 10, 20, and 30 mm. Three other corresponding sets are required at larger sizes specified on the instruction sheet (see Web site).

The exercise in Section 4.3 requires special paper sheets with dimensions as given on the instruction sheet.

The visual search task requires plotter sheets with a scatter of punctuation symbols with different sizes of search area, different sizes of search target, and either one or two targets. There should be ten examples of each generated from different random numbers in each case to vary the scatter. Master examples are available on the Web site.

For the decision-making task, a set of nine stiff cards is needed. One displays the master reference line, which is 5 mm wide and 100 mm long, in black ink on a white background. The other eight cards consist of two sets of four. One set has similar lines but two of length 98 mm and positioned slightly left or right of the vertical centreline, and at different heights between the horizontal centreline and the top and bottom edges, and two of line length 102 mm. The other set has similar lines of 96 mm and 104 mm length. For help, see the Web site. In addition, it is necessary to have a screen behind which the instructor stands and presents the cards in balanced but randomised orders by holding them over the top edge of the screen.

4.1 HUMAN DISCRIMINABILITY

OBJECTIVES

- To examine the time required to make discriminations in relation to Crossman's Confusion function (C)
- To examine the suitability of Crossman's C for this task

APPARATUS

Stopwatches.

Packs of 50-mm-square cards as described in Table 4.1. Each has 20 black discs of the standard size diameter and 20 black discs of unknown diameter.

TABLE 4.1
Dimensions and specifications of card packs

Pack No.	Standard diameter (mm)	Unknown diameter (mm)	Difference (%)	Crossman's C
W1	30	36.0	20	12.6
W2	30	34.5	15	16.5
W3	30	33.0	10	24.2
X1	20	24.0	20	12.6
X2	20	23.0	15	16.5
X3	20	22.0	10	24.2
Y1	10	12.0	20	12.6
Y2	10	11.5	15	16.5
Y3	10	11.0	10	24.2

TECHNICAL BACKGROUND

Attempts have been made over the years to establish a relationship between the time required to make discriminations and task difficulty in terms of the information-processing requirements of the discrimination required. An early approach was that of Weber, who thought that the difficulty depended only on relative magnitudes when judging between two different stimuli. Somewhat later, Fechner suggested that there was also an effect of absolute size. More recently, Stevens devised his power law, and then Crossman developed his Confusion function (Crossman 1955) (see following text).

Ergonomists need to have a good understanding of such demands on workers for a variety of situations, wherever human discrimination is required. The more difficult the task, the longer it will take to perform and the greater will be the probability of errors. These will result in impaired system performance, reduced quality and quantity of output, and the occurrence of accidents. Better understanding of these tasks means that these shortcomings can be obviated or reduced at the design stage before the system goes into operation, and Crossman's C is the type of measure that is needed to evaluate them beforehand, using the following equation:

$$\text{Crossman's C} = \frac{1}{\text{logUnknownDia} - \text{logStandardDia}}$$

PROCEDURE

1. **Participants:** Split into groups of two.
2. **Sorting:** Each participant has to sort each pack into two piles, standard and unknown, as quickly as possible, holding the pack in a single pile, face-up. The other participant times the sort and records it in Table 4.2.

TABLE 4.2
Times for sorting packs of circular discs

Part.	W1		W2		W3		X1		X2		X3		Y1		Y2		Y3	
no.	1st	2nd	1st	2nd	1st	2nd	1st	2nd	1st	2nd	1st	2nd	1st	2nd	1st	2nd	1st	2nd
1																		
2																		
3																		
4																		
5																		
6																		
7																		
8																		
9																		
10																		
11																		
12																		
13																		
14																		
15																		
16																		
17																		
18																		
19																		
20																		

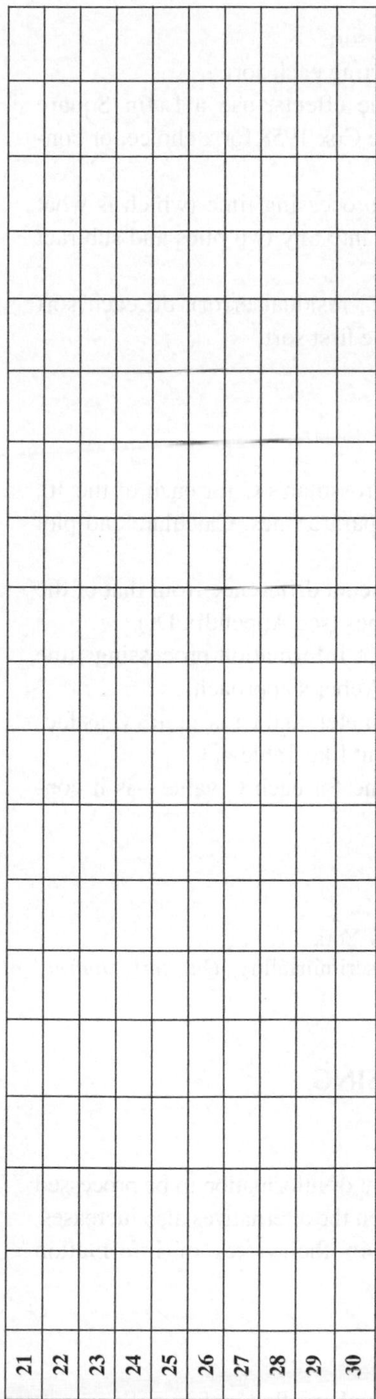

   - *Note:* Perfect performance is required; that is, if any are incorrectly classified, the sort is repeated.
   - Do a practice sort before the first timed sort.
   - Shuffle the cards thoroughly before starting each sort.
3. **Design:** To minimise learning and fatigue effects, use a Latin Square Design to allocate participants to packs (see Cox 1958 for a choice, or consult Appendix VII).
4. **Movement Time:** To get the information-processing time (which is what we need), get the time for dealing the pack into any two piles and subtract it from all the times.
5. **Replication:** To measure repeatability (i.e., residual error), do each sort twice. Use the reverse order compared to the first sort.

REQUIREMENTS

1. A long laboratory report.
2. Plot information-processing time against Crossman's C for each of the 10, 20, and 30 mm diameters, that is, three separate lines. Calculate and plot the regression line for each.
3. Test the slope of the 10 mm line for a significant difference from that of the 20 and 30 mm lines, using the *t*-test for slopes (see Appendix IX).
4. For comparison with the graphs for C, plot information-processing time against percentage difference, to examine Weber's approach.
5. Do an ANOVA with Participants, Pack Diameter, and Crossman's C as factors, and present the results in a table laid out like Table 4.3.
6. Calculate the coefficient of variation of time for each C value—is it constant or does it increase with C?

REFERENCES

Cox, D.R., 1958, *Planning of Experiments*, Wiley, New York.

Crossman, E.R.F.W., 1955, The measurement of discriminability, *Quarterly Journal of Experimental Psychology*, 7, 176–195.

4.2 CENTRAL INFORMATION PROCESSING

OBJECTIVES

- To examine the hypothesis that, as the amount of information to be processed increases, the time required to choose between the alternatives also increases
- To see differences between participants in their rates of information processing

APPARATUS

For each group, one pack of playing cards less jokers, that is, 52 in all

One stopwatch per group

TABLE 4.3
Format for ANOVA table, and mean squares to use for correct mean square ratios

Factor	Degrees of freedom (d.f.)	Sum of squares	Mean square	Mean square ratio	Significance (e.g., $p < 0.001$)
P	$n - 1$		1	1/8	
D	2		2	2/4	
C	2		3	3/5	
P*D	$2*(n - 1)$		4	4/8	
P*C	$2*(n -1)$		5	5/8	
D*C	4		6	6/7	
P*D*C	$4*(n - 1)$		7	7/8	
Residual			8		
Total	$18n - 1$				

Note: Depending on the version, the computer package may assume that all main effects are fixed effects, whereas Participants is a random effect (i.e., we cannot choose the ability level), so some mean square ratios presented may be wrong for this application, especially in older versions. Check by calculating the ratios presented in the table.

Get significance levels from tables such as Murdoch and Barnes.

*Elsewhere in your report, use *, **, ***, and **** for p < 0.05, p < 0.01, p < 0.001, and p < 0.0001, but not in this table.*

(P = Participants, D = Pack Diameter, C = Crossman's C)

TECHNICAL BACKGROUND

Hick (1952) showed that the time to process information increases directly with the amount of information to be processed when the latter is measured in bits. Crossman (1953) applied this to a series of card-sorting tasks where the amount of information was varied by sorting different combinations of cards. The results gave a very clear picture, and with a task that is easy to perform in the laboratory or classroom.

Ergonomists need to have a good understanding of such demands on workers for situations such as those that exist in control room tasks. Swain and Guttman (1983) showed how the time required to perform a series of such tasks was more than the time available before the system went out of control. The aim should be to anticipate these problems beforehand by having an adequate means to predict the time requirements, and to reduce the opportunities for error.

PROCEDURE

1. **Participants:** Split up into groups of two.
2. **Operations:** Shuffle the cards thoroughly. Then, one participant sorts the cards into various deals as in Table 4.4, *working as fast as possible*, placing the cards from left to right in the order given. The other participant takes the time. To reduce fatigue, alternate the operations between the partners.

TABLE 4.4
Descriptions and information levels of deals

Deal no.	No. of classes	Description of deals (orders on desk left to right as below)	H (bit)
1	—	Four equal piles of anything	0
2	2	Red suits/black suits	1.00
3	3	All pictures/red plain/black plain[a]	1.55
4	4	Hearts/clubs/diamonds/spades	2.00
5	6	Red pictures/black pictures/HN/CN/DN/SN[b]	2.55
6	8	Ace to 6 by suits/7 to King by suits[c]	3.00
7	13	Ace/2/3/4/5/6/7/8/9/10/J/Q/K	3.70
8	26	Deal 7 in red/deal 7 in black	4.70

[a] Pictures = Jack, queen, and king.
[b] HN = Heart's numbers, CN = Club's numbers, etc.
[c] Use the same order of suits as in Deal 5, that is, H-C-D-S/H-C-D-S.

Note:
Do an untimed practice deal first to get into a rhythm.
Correct sorting errors immediately, during the deal.
Put one card on the pile at a time.

3. **Face-up or Face-down:** Even-numbered participants do face-down first, whereas odd numbers do face-up first. Then, use the reverse order for the second set of deals. Record your results in Tables 4.5 and 4.6, respectively.
4. **Theory:** Assume perfect overlap of choice time and movement time for face-up, so this time = max (H/R, Tm), where H = bit, R = rate of information processing (bit/second), and Tm = movement time(s). For face-down the times are additive, so plotted points fall above and give two parallel lines for the results. Movement time/pile is found from D1 (i.e., no choice time) and then adjusted pro rata for number of piles.
5. **Design:** To balance approximate learning and fatigue effects, order the deals by a Latin square design (see Appendix VII).

REQUIREMENTS

1. A long laboratory report.
2. Do an ANOVA with Participants (P), H, and Face (F) as the main effects, treat P*H*F as the residual, and present these values in a table with the degree of significance for each observed Mean Square Ratio (MSR). It is a mixed effects design, so for Participants use the residual as its denominator, but for Face use P*F as MSR denominator, and for H use P*H as its denominator. For the others, use the residual. Get the levels of significance from tables such as Murdoch and Barnes. Point out the meaning of the ANOVA results (significant if $p < 0.05$).

TABLE 4.5
Face-up times (second) by deals and participants

Participant	D1	D2	D3	D4	D5	D6	D7	D8

Note: D1 = Deal 1 from Table 4.4; not the first deal for the participant in question, etc.

TABLE 4.6
Face-down times by deals and participants (second)

Participant	D1	D2	D3	D4	D5	D6	D7	D8

Note: D1 = Deal 1 from Table 4.4; not the first deal for the participant in question, etc.

3. Get the means for all ANOVA combinations and put these into tables.
4. Produce figures to graph table data, and look for interactions by plotting families of curves such as two sets of points for Face with response time versus H, and sets of points for each participant with time versus H. Join the points of each set by straight lines. Get regression lines for the figures and give the equation for each on the figure. Arrange participants in the order of increasing response times.
5. Plot total time per deal against H for each participant, slope = 1/R, and list R values for each participant for each deal in a table. Remember, 52 cards means 52 decisions.
6. Plot choice time per deal (i.e., total time – movement time) against H per deal.
7. Test the slopes of the two lines for Face, to see if they are significantly different, by means of a *t*-test as in Exercise 4.1 (see Appendix IX). Give regression equations and F, *t*, and *p* values.

REFERENCES

Crossman, E.R.F.W., 1953, Entropy and choice time: the effect of unbalance on choice response, the effect of frequency, *Quarterly Journal of Experimental Psychology*, 5, 41–51.

Hick, W.E., 1952, On the rate of gain of information, *Quarterly Journal of Experimental Psychology*, 4, 11–26.

Swain, A.D. and Guttman, H.E., 1983, *Handbook of Human Reliability with Particular Emphasis on Nuclear Power Plant Applications*, National Technical Information Service, Springfield, Virginia 22161.

4.3 MOTOR SYSTEM INFORMATION PROCESSING

OBJECTIVES

- To apply Fitts' index of difficulty (ID) to tapping movement time (MT)
- To examine the application of Fitts' ID to a type of continuous control task
- To compare ID values between the tasks
- To compare Fitts' ID with that of Welford
- To examine differences between participants on each task

APPARATUS

Tapping Task: Two sets of paper sheets with 150-mm-high rectangles of width 5, 10, and 20 mm at a pitch of 160 mm, and 10, 20, and 40 mm at a pitch of 320 mm (see Web site for masters).

Circle Tracing Task: Two sets of sheets with circles of 160 mm mean diameter and widths of 5, 10, and 20 mm, and 226.3 mm mean diameter with widths of 7.07, 14.1, and 28.3 mm (see Web site).

Stopwatch

TECHNICAL BACKGROUND

In addition to mental processing time it is also necessary to examine the time required to make limb movements such as occur when moving the hand or foot between one control and another. It is necessary to see how this relates to the distance moved, but it can be understood readily that it will take longer if the required precision of positioning at the end of the move increases. Fitts (1954) investigated this, and Fitts and Petersen (1964) looked at the combination of the two in tapping between two marked areas. They used ideas from information theory to come up with the Fitts' ID to account for the results.

$$\text{Fitts' index of difficulty (ID)} = \log_2(2A/W)$$

where A = amplitude of movement required and W = width of the target area. From these he defined movement time (MT) as

$$\text{Fitts MT} = k.\log_2(2A/W)$$

Subsequent researchers have applied the idea successfully to a wide number of applications, demonstrating its usefulness for addressing tasks of the type described (Drury 1975, Hoffman and Sheik 1994, Osinski and Drury 1995). However, Welford came to the conclusion that Fitts' ID needed a slight refinement, which he found fitted better to his data by defining movement time as follows:

$$\text{Welford MT} = k.\log_2(A/W + 0.5)$$

One of the instructive lessons to be obtained from this laboratory exercise is to compare results between his modified ID and that of Fitts.

PROCEDURE

1. **Participants:** Divide into groups of two—half the participants do tapping first, and the other participants do circle tracing first, with amplitudes and widths as given in Table 4.7.
2. **Tapping Task:** Move the stylus back and forth as fast as possible to tap alternately in left and right rectangles without interruption (amplitude = distance between centrelines).
   - Each participant does 5 initial taps (not counted) and then 20 test taps at each pitch by width.
   - Balance orders approximately using Table 4.7 to reduce order effects.
   - Get the times by stopwatch, and record them in Table 4.9.
   - To correct for errors, only count pencil dots within the rectangles.
   - Average the whole time over this number to get the mean.
3. **Circle Tracing Task:** Trace between the lines with the dominant hand going as fast as possible until five laps have been completed to get time per lap.
   - Each participant does one initial trace (not counted), and then 5 test laps at each diameter by width.

TABLE 4.7
Experimental orders for widths by task and amplitude (pitch/diameter, mm)

Subjects	Do tapping task 1st						Do circling task 2nd					
	160 pitch 1st			320 pitch 2nd			160 diameter 1st			226 diameter 2nd		
1, 7	5	10	20	20	40	10	10	5	20	28	7	14
2, 8	5	20	10	20	10	40	10	20	5	28	14	7
3, 9	10	20	5	10	20	40	5	10	20	14	7	28
	Do circling task 1st						Do tapping task 2nd					
	226 diameter 1st			160 diameter 2nd			320 pitch 1st			160 pitch 2nd		
4, 10	10	5	20	10	40	20	5	20	10	14	28	7
5, 11	20	5	10	40	10	20	20	5	10	7	14	28
6, 12	20	10	5	40	20	10	20	10	5	7	28	14

Note: For more than 12 participants, start again at the first line and continue in this fashion for the remaining participants.

- Balance orders approximately according to Table 4.7 to reduce order effects.
- Get times by stopwatch, and record them in Table 4.9.

Note:
Diameter here is to the centreline of a ring = mean diameter.
Assume amplitude here = circumference for one lap at mean diameter.
Then, ID values in all task combinations are: 5.66, 6.66, and 7.66.

4. **Residual Error:** To get a measure of the residual error for use in the ANOVA, repeat the tasks in reverse order of the rows in Table 4.7 and record the data on a second copy of Table 4.9.

REQUIREMENTS

1. A long laboratory report.
2. Check that data values are normally distributed; if not, apply an appropriate transformation.
3. Do a 3-way ANOVA with Fitts' ID, Task (T), and Participants (P) as main effects and the individual MTs for completing a tap or circle trace as the dependent variable. Present all the ANOVA data in one single table, as shown in Table 4.8. Label tasks as "Tapping" or "Circles" and IDs as the values in question, in the report. Use logs to base 2 to calculate ID.
4. Divide the results into subsections that group all the data relating to one ANOVA effect (or experimental consideration) with the ANOVA result, table of data, and figures that apply. For these, compile tables for means of all factors displayed in the ANOVA table.
5. For each task, plot MT versus Fitts' ID for each participant (using all values on one graph), and then for the means across all participants get the regression, its parameters, the correlation coefficient, and its significance.

TABLE 4.8
Format for ANOVA table, and mean squares to use for correct mean square ratios

Factor	Degrees of freedom (d.f.)	Sum of squares	Mean square	Mean square ratio	Significance (e.g., $p < 0.001$)
ID	2		1	1/5	
T	1		2	2/6	
P	n – 1		3	3/8	
ID*T	2		4	4/7	
ID*P	2*(n – 1)		5	5/8	
T*P	n–1		6	6/8	
ID*T*P	2*(n – 1)		7	7/8	
Residual			8		
Total	24n – 1				

Note: Depending on the version, your computer package may assume that all main effects are fixed effects, whereas Participants is a random effect (i.e., we cannot choose the ability level), so some mean square ratios presented may be wrong for this application, in older versions. Check by calculating the ratios presented in Table 4.8.

Get significance levels from tables such as Murdoch and Barnes.

*Elsewhere in your report, use *, **, ***, and **** for $p < 0.05$, $p < 0.01$, $p < 0.001$, and $p < 0.0001$, but not in this table.*

Renumber participants in ascending order of their overall times to show a pattern in the picture, if there is one. Do tapping tasks differ from each other, circle tasks from each other, and tapping from circling?

6. Plot the means of all participants on each of the tasks for all the data to give two regression lines on one graph in order to see any differences between them, giving the line parameters plus correlation coefficient with its significance.
7. Compare the fit of the data, averaged over participants, to Fitts' ID and Welford's ID, separately for each task.
8. Compare the rate of information processing (reciprocal of slope, i.e., R) for circles versus tapping and between participants both within and between tasks.
9. Plot families of curves to examine two-way interactions, that is, MT versus Fitts ID points for each task, MT versus Participants (in ascending order of MT, *not* number), and MT versus P for each task.
10. Calculate individual regression parameters for each participant on each task; list them in a table.

REFERENCES

Drury, C.G., 1975, Application of Fitts' Law to foot pedal design, *Human Factors*, 17, 368–373.

TABLE 4.9
Task times (s) for tapping and circle tracing tasks

Participant no.	Tapping task (per move)						Circle tracing (per lap)					
	160			320			160			226		
	5	10	20	10	20	40	5	10	20	7	14	28
1												
2												
3												
4												
5												
6												
7												
8												
9												
10												
11												
12												
13												
14												
15												
16												
17												
18												
19												
20												
21												
22												
23												
24												
25												
26												
27												
28												
29												
30												
31												
32												
33												
34												
35												

Fitts, P.M., 1954, The information capacity of the human motor system in controlling the amplitude of movement, *Journal of Experimental Psychology*, 47, 381–391.

Fitts, P.M. and Peterson, J.R., 1964, Information capacity of discrete motor responses, *Journal of Experimental Psychology*, 67, 103–112.

Hoffmann, E.R. and Sheik, I.H., 1994, Effects of varying target height in a Fitts' movement task, *Ergonomics*, 36, 1071–1088.

Osinski, C.J. and Drury, C.G., 1995, Accurate movement of two-probe components, *Ergonomics*, 38, 337–346.

Note: To find logs to the base of 2

Recall that the log of Y is the power X to which the base must be raised in order to equal Y. That is, $2^X = Y$, so if we take logs to the base 10 on both sides, we will get $X.\log_{10}2 = \log_{10}Y$. Hence, $X = (\log_{10} Y)/(\log_{10}2)$, so $X = (\log_{10} Y)/0.30103$.

4.4 VISUAL SEARCH

OBJECTIVES

- To examine the effects of different sizes of search area, target height, and number of targets on search time.
- To examine the statistical distribution of search times
- To see the effects of individual differences on search performance
- To examine the efficacy of different measures of search times, especially on measures of central tendency

APPARATUS

Stopwatches

Photocopies of eight sets of plotter search sheets with 10 in each set, with sets labelled 1 through 8 (basic details are given in Table 4.10)

TABLE 4.10
Details of the plotter sheet sets

Area	Small (25 rows * 30 columns)				Large (25 rows * 50 columns)			
No. targets	One		Two		One		Two	
Target Ht.	Small	Large	Small	Large	Small	Large	Small	Large
Set	1	2	3	4	5	6	7	8
Latin Square	E	C	A	H	B	D	G	F

Note: For experimental ordering, using random numbers from Appendix VII, put A = 3, B = 5, C = 2, D = 6, E = 1, F = 8, G = 7, and H = 4.

TECHNICAL BACKGROUND

Visual search relates directly to the issue of human reliability, in addition to its applications in visual inspection. In many situations, people are employed in monitoring tasks where they have to search a bank of displays in a continuous sweep, looking for any sign of a possibly out-of-control condition, for example, steel sheet inspection, control room work, cockpit operations, maintenance inspection, and military operations. These tasks raise issues of how big an area to search in each sweep, whether to search in a random or systematic pattern, what "size" of object will stand out from the background, what degree of success is likely, and how much time it needs.

Melloy et al. (2000) applied this material to aircraft maintenance inspection, and Drury has done extensive research, particularly in the area of industrial visual inspection (Gallwey and Drury 1986; Gallwey 2000). Among other things, Drury (1978) established some models of the process, building on early work by Engel (1977) and the work of others. Engel demonstrated that search time depends on a number of variables such that:

$$t_m = t_0.A/(N.a.P_0) \text{ seconds}$$

where t_m = mean search time

t_0 = duration of individual fixations (often assumed to average 300 milliseconds)

A = area of the field being searched

N = number of targets

a = area around the line of sight in which targets can be detected in t_0 seconds

= visual lobe size

P_0 = probability of detecting a target if it falls within "a"

Some of these can be demonstrated experimentally here.

The material viewed here consists of ten punctuation type symbols, and the "target" is a stylised left parenthesis "(", which is present at either one of two heights (H), which implies two values of "a". Similarly, there are two values of "A", that is, big and small. Each target appears either once or twice only on each sheet, and we assume that t_0 and P_0 are constants. So, effectively $t_m \propto A/(N*H)$, and differs between Participants. There are ten sheets for each combination, with the symbols spread about in a random pattern with 20% of filled spaces, and the position of the target is chosen at random. Data in the literature show that such search times are exponentially distributed, which means that they must be log-transformed to meet the normal distribution requirement for the ANOVA. Then, the antilog values of the means of these transformed times will be geometric mean search times (GMSTs). (*Note*: An alternative measure to address nonuniformly distributed data is to use the median).

The exponential distribution has the parameter λ, which in this case characterises the particular search task undertaken and, if the data really are exponentially distributed, this λ value will be the same as the reciprocal of the arithmetic mean μ. This point needs to be checked (see the attached tutorial sheet in particular). The other characteristic of the exponential distribution is that the arithmetic mean μ has the same value as the standard deviation (SD), which also needs to be checked.

It is necessary to draw a distinction between visual inspection and visual search. In visual search a "target" is always present. In visual inspection the "target" is hopefully absent most of the time, so that the searcher has to have a decision rule for when to stop searching the area. Secondly, if a "target" is found, the searcher usually has to make a judgement as to whether it should be accepted or rejected, which can be difficult especially in borderline cases. These extra features make it much more complex, so it has not been pursued here.

PROCEDURE

1. Divide into groups of two and collect the search sheets, but do not allow the searcher to see the sheet until it is time to start.
2. Then one searches each sheet in the set for the left parenthesis, working as quickly as possible, until the first target only is found.
3. When found, indicate it by pointing a finger but do not mark the sheet.
4. The other person captures and records this time to find the "target".
5. Each person searches all sheets of all sets and, to balance learning and fatigue effects, use the Latin square order given in Table 4.11. On each sheet, get the time for the first target only.
6. The ten sheets (or examples) of each combination represent ten "replications", so enter the data in that fashion for the statistical analysis. Enter all

TABLE 4.11
Orders of sets of plotter sheets for each participant

Participant nos.	Order of sets							
	1st	2nd	3rd	4th	5th	6th	7th	8th
1, 9	1	2	5	7	8	4	6	3
2, 10	2	3	4	1	6	8	5	7
3, 11	8	6	2	4	7	3	1	5
4, 12	5	4	7	6	2	1	3	8
5, 13	6	5	3	8	1	7	2	4
6, 14	4	8	1	5	3	2	7	6
7, 15	3	7	8	2	5	6	4	1
8, 16	7	1	6	3	4	5	8	2

Note: If there are more than 16 participants, start again at participant 1; that is, 17 does the same as 1 and 9, 18 does the same as 2 and 10, and so on.

Give the raw time values a label in the computer package such as "rtime". Then, for the ANOVA, get "ltime" (say) as the label where ltime = log(rtime). Use ltime to get GMSTs, rtime for medians, AMTs, and SDs.

Take antilogs of means of ltime to get GMST in seconds, not log-seconds, for each set of 10.

raw search times as found (i.e., do not enter the means or standard deviations), and enter levels as 1 or 2, not as sizes. Then, within the package, label them as "Small" or "Large", "One" or "Two", and Participants as 1,2,....

7. All participants search all the sets again but in the reverse order, to measure the residual error.

REQUIREMENTS

1. A long laboratory report.
2. Do an ANOVA with Participant (P), Area (A), Number of targets (N), and Height (H) as the main effects. To get approximately normally distributed values for the ANOVA, use log(search time). Present all factors (main effects and all interactions and the residual) in one ANOVA table, as in Table 4.12.
3. Take antilog of means of log(search time) to get geometric mean search times (GMSTs). Produce tables of GMSTs across participants for the various combinations of A, N, and H as for all the 15 factors in the ANOVA table, but structure them on the same basis as Table 4.10.
4. Draw graphs for the significant ANOVA factors, and comment on what they show. For two-way interactions, plot two sets of points for the dependent variable, with the independent variable on the horizontal axis as usual. For three-way plot, four sets of points (e.g., 1 target @ large, 2 targets @ large, 1 target @ small, and 2 targets @ small), with the other on the horizontal axis (e.g., small area, large area). Use descriptive labels for levels of variables (e.g., small and large) rather than coded terms.
5. Each participant gets arithmetic mean time (AMT) and standard deviation (SD) of time for self, and plots t versus A/(N*H), gets regression with correlation and its significance, and plots it. Use A values of 25*30 and 25*50, and H values of 5.3 and 8.9. Remember that they are fixed effects.
6. Plot average median time, AMT, and GMST averaged over all participants versus each value of A/(N*H), get regression with correlation and its significance for each, and plot them on the one graph. Comment on these measures of central tendency in the light of the data.
7. Plot cumulative proportion found versus cumulative time (use intervals of, say, 3, 4, or 5 seconds) for each condition of A/(N*H), one graph for each area with four sets of points on each, calculated over all data.
8. Plot two graphs of time versus participants (in ascending order of times, *not* participant number), one for each area, with four separate sets of points for target height (H)*number of targets (N).
9. Check if the cumulative distribution of times is exponential: Get the λ value for each combination of A, N, and H across all participants using linear regression (see tutorial). Also, see if λ = reciprocal of AMT, and if AMT is approximately equal to SD. Calculate correlation coefficients for these two combinations. Give all the data for these calculations.

TABLE 4.12
Format for the ANOVA table, and identification of the mean square values to be used in calculating the correct mean square ratios

Factor	Degrees of freedom	Sum of squares	Mean square	Mean square ratio	Significance (e.g., p < 0.001)
P	*n* – 1		1	1/16	
A	1		2	2/5	
N	1		3	3/6	
H	1		4	4/7	
P*A	*n* – 1		5	5/16	
P*N	*n* – 1		6	6/16	
P*H	*n* – 1		7	7/16	
A*N	1		8	8/11	
A*H	1		9	9/12	
N*H	1		10	10/13	
P*A*N	*n* –1		11	11/16	
P*A*H	*n* – 1		12	12/16	
P*N*H	*n* – 1		13	13/16	
A*N*H	1		14	14/15	
P*A*N*H	*n* – 1		15	15/16	
Residual			16		
Total	80*n* – 1				

Note: Replace n by the appropriate number of participants when putting it in the report.
*Use N.S. if not significant, code elsewhere as *, **, ***, and **** for <0.05, <0.01, <0.001, and <0.0001 probability of getting the result purely by chance, but use figures in this table.*
Some computer packages assume that all main effects are fixed, that is, that we have exact levels of each. This is obviously not so for Participants.
Look up significance values in tables such as Murdoch and Barnes.

REFERENCES

Drury, C.G., 1978, Integrating human factors in statistical process control, *Human Factors*, 20, 561–570.

Engel, F.L., 1977, Visual conspicuity, visual search and fixation tendencies of the eye, *Vision Research*, 17, 95–108.

Gallwey, T.J., 2000, Evaluation and control of industrial inspection, In *Ergonomics Guidelines and Problem Solving* (Vol. 1), Mital, A., Kilbom, A., and Kumar, S. (Eds.), Elsevier, Amsterdam.

Gallwey, T.J. and Drury, C.G., 1986, Task complexity in visual inspection, *Human Factors*, 28, 595–606.

Melloy, B.J., Harris, J.M. and Gramopadhye, A.K., 2000, Predicting the accuracy of visual search performance in the structural inspection of aircraft, *International Journal of Industrial Ergonomics*, 26, 277–283.

TUTORIAL ON VISUAL SEARCH DATA

Use the data collected in this experiment to explore some of its characteristics. For convenience this is broken into three steps.

1. **Exponential distribution:** If cumulative search times are exponentially distributed, we will get this function.

$$F(t) = 1 - e^{-\lambda t}$$

We can rewrite this as

$$1 - F(t) = e^{-\lambda t}$$

If we take the reciprocal of both sides, we get

$$\frac{1}{(1 - F(t))} = e^{\lambda t}$$

Now take natural logs (to the base e) of both sides to get

$$\ln \frac{1}{(1 - F(t))} = \lambda t$$

If the left-hand term is called y, we can rearrange the equation as:

$$y = 0 + \lambda t$$

which is the equation for a straight line with constant = 0 and slope = λ. In this case, λ = the visual search parameter.

2. **Use regression** to get the line of best fit, and hence, its slope (λ). Use a calculator routine or a software package, or set up a table with columns for y_i, x_i, $x_i y_i$, and $(x_i)^2$ and then sum these columns. The straight line equation in general form is

$$y = a + b.x$$

$$\text{where a} = \frac{\Sigma y_i - b\Sigma x_i}{n} \text{ and b} = \frac{n\Sigma x_i y_i - \Sigma x_i \Sigma y_i}{n\Sigma x_i^2 - [\Sigma x_i]^2}$$

In this application we have:

$$y_i = \ln \frac{1}{\left(1 - F(t_i)\right)}$$

with $x_i = t_i$ or cumulative time in each interval,

and n = number of intervals.

3. **Compare data from the visual search experiment.** Use it to get the following:

Arithmetic mean times (AMT) and standard deviations (SDs): Calculate these from the times for each set of sheets (i.e., one set of ten sheets) for your data.
Note:

$$SD = \{1/(n - 1)[\Sigma x_i^2 - ((\Sigma x_i)^2)/n]\}^{0.5}$$

Median times (MT): Obtain these from the times for each set of sheets from your data (half the values are above it and half are below it, but for an even number of values, take the mean of the two middle ones), that is, one per set.

Geometric mean STs (GMST): Calculate this; that is, take logs, sum (Σ) logs, get their mean, and then get the antilog value of that mean, and one value per set of ten sheets. Use these values (seconds) also in ANOVA, calculations, etc.

MTs, AMTs, and GMSTs comparison: For each of the same sets, are they approximately the same across participants? Are there any other patterns or trends in these data?

λ values: Calculate by using regression, etc., given earlier, for the data of all participants, for each set of sheets. Share them out among the members of the class. Strictly speaking, the λ values will be different for each person, so they should be treated as such if that is possible. However, with only ten values per participant per condition, that is insufficient to establish any real cumulative data.

Plot of cumulative distribution: This is the cumulative proportion of targets found versus the cumulative search times (data from all participants) to get a picture of the cumulative distribution of search times. What proportion of the targets was found from 0 to 5 seconds (say), then between 0 and 10, then total 0 to 15, etc.? Plot these at the end of each interval, i.e., at 5, 10, 15, etc. Do this for each combination of A, N, and H.

Check on the cumulative exponential distribution: In other words:

Does λ = the reciprocal of the arithmetic mean time (AMT or μ) approximately?

Does the arithmetic mean time (AMT) = the standard deviation value (approximately)? If so, then it is very likely that the distribution is exponential.

4.5 DECISION MAKING (TSD)

OBJECTIVES

- To see if the Theory of Signal Detection (TSD) gives independent measures of sensitivity and bias
- To see if these are affected by task difficulty, and hence, provision of a reference standard
- To compare parametric and nonparametric measures of the same data

APPARATUS

Master card with a black line 5 mm wide by 100 mm long (see Web site)
Other cards with 5 mm lines of unknown length (Web site)
Board on a bench to screen the instructor from the students
Random order plan for presentation of the cards by the instructor

TECHNICAL BACKGROUND

A common area of human reliability failures is that of decision making. There are two components, the difficulty of detecting a difference between two stimuli, and the personal bias (or organisational pressure) to lean towards acceptance or rejection in the decision. In traditional perception-type tasks (e.g., auditory perception levels), these are inextricably mixed together. However, in TSD Tanner and Swets (1954) envisaged the task as trying to separate signal from noise in a communications task. They depicted these stimuli as two overlapping normal distributions where the Human Observer (H.O.) sets some criterion value of the evidence variable whereby above it H.O. reports signal and below it reports noise. The distance between the means of the distributions represents the detectability of the task (d') and the position of the criterion value (X_c) gives the bias of the H.O., represented by the ratio of the two probabilities at the X_c value (called β). Thus, they devised a system to measure these two factors separately. If the theory works fully, there should be no effect of one on the other, that is, they should be mutually orthogonal. For more details, see Green and Swets (1974) and McNicol (1972). It has been found to work in a large variety of different applications and so can be very helpful to the ergonomist, but there are some alternative views (Craig 1977; Long and Waag 1981; See et al. 1997).

It has been recognised for a long time that judgement decisions (in particular, for inspection) will be made more accurately, and with finer gradations, if they are made relative to a reference standard rather than on an absolute basis (Van Cott and Kincade, 1972). The opportunity is taken here to test this as a simple application of the TSD theory. Providing a reference standard should increase the d' value, but should not effect any change in the β or bias value. Similarly, if the task is easy, the d' value should be larger than if the task were difficult, but it should not change

TABLE 4.13
With reference standard (mark "L" if you think the line is longer, "S" if shorter)

No.	Mark	No.	Mark	No.	Mark	No.	Mark	No.	Mark
1		21		41		61		81	
2		22		42		62		82	
3		23		43		63		83	
4		24		44		64		84	
5		25		45		65		85	
6		26		46		66		86	
7		27		47		67		87	
8		28		48		68		88	
9		29		49		69		89	
10		30		50		70		90	
11		31		51		71		91	
12		32		52		72		92	
13		33		53		73		93	
14		34		54		74		94	
15		35		55		75		95	
16		36		56		76		96	
17		37		57		77		97	
18		38		58		78		98	
19		39		59		79		99	
20		40		60		80		100	

the β value, and this is also tested here. At the same time, the experiment provides an opportunity to try to compare the efficacy of these parametric TSD measures against nonparametric measures (Hodos, 1970), because TSD has been dismissed on occasions on the grounds that in the "real world", the distributions will not be normally distributed.

PROCEDURE

1. The test is run on a classroom basis in two parts, with the students recording their responses individually:

    Condition 1: The master card is displayed horizontally for duration *X*.

    Condition 2: There is no master card available.

    Of course, it would be preferable for half the class to do these in the reverse order but the experiment takes a short time, so there is not much of a fatigue effect. However, it may mean that the performance in Condition 2 is better than it would otherwise have been. That is a matter for discussion afterwards. Purists may choose to run it twice to achieve that degree of balance of orders.

2. The unknown cards are displayed horizontally on the class side of the screen in random order for about 5 seconds, and the trial number is called out. There are 100 trials each time.
3. For each card, the students decide if the line is shorter (S) or longer (L) than the line on the master card, and record their decision in Table 4.13. *Note:* Responses of "same" or "don't know" are not permitted.
4. Repeat the experiment for Condition 2, and record these decisions in Table 4.14.

ANALYSIS

The data are sorted and tallied to draw up decision matrices for each participant for each case, using Tables 4.15 and 4.16. These are used to calculate the P_1 and P_2 values, and these are then used to get the corresponding z and y values from the inverse normal distribution table (Appendix X). From these, d' and Beta values are

TABLE 4.14
Without reference standard (mark "L" if you think the line is longer, "S" if shorter)

No.	Mark	No.	Mark	No.	Mark	No.	Mark	No.	Mark
1		21		41		61		81	
2		22		42		62		82	
3		23		43		63		83	
4		24		44		64		84	
5		25		45		65		85	
6		26		46		66		86	
7		27		47		67		87	
8		28		48		68		88	
9		29		49		69		89	
10		30		50		70		90	
11		31		51		71		91	
12		32		52		72		92	
13		33		53		73		93	
14		34		54		74		94	
15		35		55		75		95	
16		36		56		76		96	
17		37		57		77		97	
18		38		58		78		98	
19		39		59		79		99	
20		40		60		80		100	

Note: Master line = 100 mm.
Easy: Short = 96 mm, long = 104 mm; **Difficult:** Short = 98 mm, long = 102 mm.

TABLE 4.15
Tally of scores with reference standard

Easy			Difficult		
Response	**Actual long**	**Actual short**	**Response**	**Actual long**	**Actual short**
Long			**Long**		
Short			**Short**		
Totals			**Totals**		

TABLE 4.16
Tally of scores without reference standard

Easy			Difficult		
Response	**Actual long**	**Actual short**	**Response**	**Actual long**	**Actual short**
Long			**Long**		
Short			**Short**		
Totals			**Totals**		

Note: If one of the cells above is zero, we will get z = infinity. So, assume that one response in the other cell of the column could have gone the other way, then split it as 0.5 to the zero cell and 0.5 to the other one.

P_2 = (No. of Long Responses | Actual Long)/ Total Actual Long =

P_1 = (No. of Short Responses | Actual Short)/ Total Actual Short =

β = (y ordinate for P_2)/(y ordinate for P_1) =

calculated for each person; from P_1 and P_2, calculations are made to get A_G and B'_H values (see formulae in following text), and ROC curves are drawn where:

P_2 = (No. of Long Responses | Actual Long)/ Total Actual Long
P_1 = (No. of Short Responses | Actual Short)/ Total Actual Short

REQUIREMENTS

1. A long laboratory report.
2. Do two ANOVAs on the full set of data with main effects of Condition, Difficulty, and Participants—one with *d'* as the dependent variable, and one with Beta as the dependent variable. *Note:* Because the experiment is done only once (i.e., there is no replication), there is no residual term, so the three-way interaction is used as the residual. Also, this is a so-called "mixed" design (i.e., it has both fixed and random effects), so some of the mean square ratios displayed may be wrong. The correct ratios are Condition – C/(C*P); Difficulty–D/(D*P); all others use the new Residual as the denominator.

3. Because we do not know the sampling distributions for d' and Beta, and they are almost certainly not normal (needed for ANOVA), we need to check on the validity of the ANOVA by doing Friedman tests ignoring Participants (because it can only do a two-way at most).
4. Test for correlation between d' and Beta and for A_G and B'_H in each case. Are they significant?
5. Test for correlations between d' and A_G. Then do the same for Beta and B'_H for each case. Are they significant?
6. Plot an ROC curve for all four cases, with all participants on the same plot if there is enough space, with different symbols for each and only the plots of each set joined up with straight lines.
7. Arranging participants in increasing order of d' for Easy with Reference standard, plot on the one graph d' against Participants with one set of points for Easy and Reference standard (ER), another for Difficult plus Reference (DR), one for Easy Without standard (EW), and one for Difficult Without standard (DW). Use different symbols for each of ER, DR, EW, and DW, and join up the points of each set.
8. Draw the corresponding graph to 7 for Beta, A_G, and B'_H.
9. Plot one graph of d' versus A_G with separate sets of points for each of the four cases (ER, DR, EW, and DW), with the points of each set joined by straight lines.
10. Plot one graph of Beta versus B'_H for each case, as for 9.
11. Provide a table of number of S given *s*, S given *n*, N given *s*, and N given *n*, plus P_1, P_2, z_1 and z_2, and the calculations, in an Appendix.
12. Draw up one table summarising all the correlation results, one for d' and β (with Reference and Without) for Easy and Difficult, another for A_G and B'_H in the same way, in a tree structure.

REFERENCES

Craig, A., 1977, Broadbent and Gregory revisited: vigilance and statistical decision, *Human Factors*, 19, 25–36.

Green, D.M. and Swets, J.A., 1974, *Signal Detection Theory and Psychophysics*, Wiley, New York.

Hodos, W., 1970, Nonparametric index of response bias for use in detection and recognition experiments, *Psychological Bulletin*, 74, 351–354.

Long, G.M. and Waag, W.L., 1981, Limitations on the practical applicability of d' and Beta measures, *Human Factors*, 23, 285–290

McNicol, D., 1972, A *Primer of Signal Detection*, Allen and Unwin, Sydney.

See, J.E, Warm, J.S., Dember, W.N. and Howe, S.R., 1997, Vigilance and signal detection theory: an empirical evaluation of five measures of response bias, *Human Factors*, 39, 14–29.

Tanner, W.P. and Swets, J.A., 1954, A decision-making theory of visual detection, *Psychological Review*, 61, 401–409.

Van Cott, H.P. and Kincade, R.G., 1972, *Human Engineering Guide to Equipment Design* (Rev. ed.), Superintendent of Documents, U.S. Government Printing Office, Washington D.C. 20402.

For the parametric case

Look up tables of the inverse normal distribution in Appendix X.

Remember, the left-hand distribution represents "noise" (i.e., here short lines displayed) and the right-hand one represents "signal" (i.e., here long lines displayed), and the criterion value X_c is somewhere in between these, but can be to the left of the noise mean or to the right of the signal mean. Sketch out the distributions and shade in P_1 and P_2 to see where X_c is relative to the means of the two distributions, and hence, how to combine z_1 and z_2 to get d' for each case.

Note: Remember that P_2 is for the right-hand tail, and P_1 is for the left-hand tail, whereas the inverse normal table always gives values for the left-hand tail only. So, to get the correct sign for z_2, look up $1 - P_2$ so that, if P_2 is greater than 0.5, the z_2 value will be negative; that is, the X_c line will be to the left of the mean of the signal distribution. Similarly, z_1 will be negative if P_1 is less than 0.5.

If X_c lies to the left of the noise mean,

z_2 will be negative and more negative than z_1.

Then d = absolute value of $(z_2 - z_1)$.

If X_c lies between the two means,

z_2 will be negative and z_1 will be positive.

Then d = absolute value of $(z_2) + z_1$.

If X_c lies to the right of the signal mean,

z_2 will be positive but smaller than z_1 (also positive).

Then $d = z_1 - z_2$.

For the nonparametric case

Here, it is usual to plot the Receiver Operating Characteristic (ROC) curve, where $P(S|s)$ is plotted against $P(S|n)$ on a unit square. The plotted points are joined by a series of lines from (0,0) to (1.0,1.0), and the greater the area under this curve (A_G), the easier the task is for H.O.; so this is a measure of discriminability.

$$A_G = \Sigma \text{ of trapezoidal areas under the ROC curve}$$

However, it is also the case that points lying below the negative diagonal represent a strict criterion and those lying above it constitute a lax criterion, and Hodos (1970) used this to define his nonparametric measure of H.O. bias (B'_H) as follows:

$$B'_H = 1 - \frac{x(1-x)}{y(1-y)} \quad \text{for the strict side}$$

and $$B'_H = \frac{y(1-y)}{x(1-x)} - 1 \quad \text{for the lax side}$$

where x = the value of P(S|n) for the point concerned
and y = the value of P(S|s) for the point concerned.

Calculate values of each variable for each participant in each condition.

TUTORIAL ON THEORY OF SIGNAL DETECTION

1. A detection experiment has targets on the left-hand side and right-hand side of the fixation point. Targets on the left-hand side count as signals, and the H.O. is asked to give a response of Definitely Signal, Maybe Signal, Maybe Noise, or Definitely Noise. Their decision results are shown in Table 4.17.

   Calculate d' and β for each case. Check to see whether or not parametric TSD is acceptable; that is, is d' constant, does β reduce when going from strict to lax, and do we get a straight line when we plot the ROC curve using z values for probability of signal given signal and for probability of signal given noise? Repeat the above using nonparametric TSD; that is, plot the ROC curve and calculate the area under the curve and B'_H.
2. Two inspectors looked for possible defects on a set of car mirrors and had to classify their decisions as Definitely Reject, Maybe Reject, Maybe Accept, and Definitely Accept. They tabulated the decisions in Table 4.18.

TABLE 4.17
Data for TSD problem

Responses	Actually left	Actually right
Definitely Signal (DS)	9	1
Maybe Signal (MS)	7	4
Maybe Noise (MN)	4	6
Definitely Noise (DN)	0	9
Totals	20	20

Hint: Where there is a zero cell, it is assumed that the response could have gone either way; so one can add 0.5 to that cell and subtract 0.5 from the cell above.

TABLE 4.18
Data for mirror inspection task

Inspector	Inspector A		Inspector B	
Real Condition	Good	Reject	Good	Reject
Definitely Reject	17	66	29	82
Probably Reject	14	14	11	3
Probably Accept	9	6	11	2
Definitely Accept	60	14	49	13

Analyse these data using both types of TSD analysis to see if there is any difference in the sensitivity or bias (or both) of these people, and comment on the meaning of these findings. Suggest improvements you would make to improve performance on this job.

3. An inspector has these responses:
   for Signals: DS 7, MS 12, MN 11, DN 2.
   for Noise: DN 16, PN 10, PS 5, DS 1.
   Use both parametric and nonparametric TSD to analyse them.

5 Physiological Issues

In devising these exercises, one of the criteria has been that they should not require any invasive techniques. This imposes some serious limitations, but also ensures a reasonable level of safety and therefore reduces the difficulties with getting acceptance from the Human Ethics Committee. Unfortunately, some of them require rather expensive equipment, but there are several that can be carried out fairly simply.

Harmonising Education and Training Programmes for Ergonomics Professionals (HETPEP) requires a basic knowledge of anatomy and physiology, which should provide an adequate basis for understanding the issues involved. Ideally, one would like to have a laboratory environment, where temperature, humidity, and air velocity can be controlled within close limits, but that would be hoping for rather too much for most people in terms of expense. Hence, some of the experiments are more in the nature of demonstrations, but they can still show fairly well what happens to people, and provide a basis for thinking and reading more on the subject.

For many people, ergonomics is synonymous with back problems; so, some work is included on this topic, with the intention that it be linked to human anatomy and physiology to explain some of the mechanisms involved. One of them provides a link with biomechanics as it is felt that this aspect needs to be approached, even though the treatment in the lecture material will be fairly brief in many cases. Although the complex issues of kinematics and mathematical analyses are usually not covered in an ergonomics course, the principles that can be shown provide a valuable insight into this important aspect of ergonomics.

Postural issues have been to the fore for a long time, and both simple and more complex methods are covered. Rapid Upper Limb Assessment (RULA) and maybe Rapid Entire Body Assessment (REBA) are in wide use but there is not much agreement on more detailed methods of analysis. However, industrial experience suggests that Drury's approach gives a lot of useful information. Therefore, the idea has been to combine these in a fashion that should be eminently practicable for most ergonomists. In effect, these and the anthropometry work are reinforced by the Assignment on a Workplace System Evaluation (Chapter 6, Section 6.4), which is in the nature of a field study.

Increasingly, the LabVIEW suite of programs has become widely used in much ergonomics work, as it can be used to design and present the experimental interface, to control the experiment, and to collect data online. Introduction to its use is highly desirable, but many students find the learning curve quite steep and, therefore, rather time consuming. Presumably, in time, off-the-shelf add-ons will become available for the more common ergonomics tasks; however, in their absence, it is not deemed appropriate to include such usage here. Certainly, in departments that use this software a great deal, specially developed routines will be available for immediate student use, but this is not likely to be the case in general.

PARTICULAR EQUIPMENT NEEDS

The Hot Box used in Section 5.7 is an enclosure consisting of a steel frame with Perspex sides and top to provide a space slightly bigger than that of an average toilet, with a drip tray covering the floor and an urn of boiling water (see Web site for an example).

The boxes used in other experiments are about the size of a briefcase and filled with sand and steel bars to adjust to specified weights of 100, 150, 200, 250, and 300 N, with two of each to facilitate testing of several participants at once (see Web site).

To estimate task dimensions and distances (e.g., in lifting investigations), a background must be provided for taking side-on photographs. Thus, wall boards are required, painted white with black grid lines at 100 mm spacing. They should be made from 2.4 × 1.2 m plywood sheets to facilitate transport to workplaces.

The anthropometry exercise requires a special wooden seat, which can be made from plywood, on a wooden frame. The seat should be flat and supplemented by a vertical block of 0.5 m height that can be pressed against the back of the participant's buttocks, and it needs a side rail on the seat to locate this block parallel to the front edge of the seat when measuring the buttock–knee length, for example (see Web site for drawings).

5.1 FORCE–ANGLE RELATIONSHIPS

OBJECTIVES

- To examine the change in muscle strength with change in joint angle
- To examine individual differences in strength
- To compare the effects of range of motion on strength

APPARATUS

Grip strength tester
Manual goniometer
Anthropometer (e.g., Holtain)
Balance scale

TECHNICAL BACKGROUND

Many manual tasks require that persons exert muscular forces with their upper arm joints in a variety of articulations but, at the same time, good design dictates that these forces should not be greater than 20–30% of maximum voluntary contraction (MVC) if exerted continuously. Because the MVC reduces as the joint is rotated away from the neutral position, it is necessary to have a good idea of the nature of this change with joint angle. This is part of the process for avoiding overstressing of the muscles or overloading of the tendons and soft tissues (Lin et al., 1994).

At the same time, good practice requires that tasks be designed so that all activities take place "with the joints at about the midpoint of their range of movement" (Appendix III, No. 7). One of the key joints is the wrist, and this exercise provides an opportunity to become familiar with the three types of wrist movements and their ranges of motion. For details of these measurements, see Norkin and White (1995). Together, these two sets of measurements build a basis for understanding the major mechanisms of Repetitive Strain Injuries and postural problems in general. They also provide a good understanding of the anatomy of the hand–arm system.

PROCEDURE

1. Split into groups of three: one as participant, one for measuring, and one as scorer. Then change.
2. Measure and record the stature (mm) and body mass (kg) of each participant as in Table 5.2.
3. Measure the range of motion (ROM) of the wrist using the goniometer as follows:
   (a) **Wrist flexion**—Position the participant in a sitting position next to a supporting surface. The upper arm is abducted to 90° and the elbow is flexed to 90°. The forearm is positioned in pronation, that is, so that the palm of the hand faces the ground. The forearm rests on the supporting surface, but the hand is free to move. Avoid radial/ulnar deviation and flexion of the fingers.
   To align the goniometer:
   – Centre the fulcrum of the goniometer over the lateral aspect of the wrist over the triquetrum.
   – Align the proximal arm with the lateral midline of the ulna, using the olecranon and ulnar styloid processes for reference.
   – Align the distal arm with the lateral midline of the fifth metacarpal.
   (b) **Wrist extension**—Do the same as for flexion, but avoid extension of the fingers.
   To align the goniometer: Do the same as for wrist flexion.
   (c) **Radial/ulnar deviation**—The testing position is the same as for flexion. Stabilise the distal ends of the radius and ulna to prevent pronation/supination and elbow flexion beyond 90°.
   To align the goniometer:
   – Centre the fulcrum of the goniometer over the middle of the dorsal aspect of the wrist over the capitate.
   – Align the proximal arm with the dorsal midline of the forearm, using the lateral epicondyle of the humerus for reference.
   – Align the distal arm with the dorsal midline of the third metacarpal. Do not use the third phalanx for reference.
4. Measure the participant's hand from the proximal interphalangeal joint of the third metacarpal to the centre of the palm, and adjust the handles of the dynamometer appropriately. Measure the maximum grip strength of the

TABLE 5.1
Random sequences of joint angle proportions (%) by participant

Sequence	Participant nos.				
	1, 6	2, 7	3, 8	4, 9	5, 10
First	100	25	0	50	75
Second	25	0	75	100	50
Third	0	75	50	25	100
Fourth	50	100	25	75	0
Fifth	75	50	100	0	25

Note: For more participants: Participant 11 is the same as 1 and 6, 12 the same as 2 and 7, and so on. *Use the same sequence for all measurements on a particular participant.*

participant at intervals of 25% of the ROM, using the grip strength analyser according to the sequence in Table 5.1, derived from a Latin square.

5. Start the test program on the software, if available. Position the participant's wrist at the appropriate angle. Instruct the participant to grip the handles using maximum force on hearing a signal, and to maintain this level of force over 6 seconds, until another signal is heard. Record the value in Table 5.2. Allow a rest period of 1 minute. Repeat this for the other intervals and for each type of motion.

REQUIREMENTS

1. A short laboratory report.
2. Plot graphs of the force–angle relationships of all the participants over the full range of flexion/extension and ulna/radial deviation.
3. Examine statistically (*t*-tests) how your data values compare with the data given in Van Cott and Kincade (1972), and with that from Pheasant and Haslegrave (2006).
4. Do a two-way ANOVA on the effects of participants and %ROM on the force produced.
5. Examine the data for any relationship between strength and stature and/or body mass on the one hand, and between joint rotation and stature and/or body mass on the other.

REFERENCES

Lin, M., Radwin, R.G., and Snook, S.H., 1994, Development of a relative discomfort profile for repetitive wrist motions and exertions, *Proceedings of the 12th Triennial Congress of the International Ergonomics Association*, Toronto, August 15–19, 2, 219–221.

Norkin, C.C. and White, D.J., 1995, *Measurement of Joint Motion* (2nd ed.), Davis, Philadelphia.

Pheasant, S. and Haslegrave, C.M., 2006, *Bodyspace: Anthropometry, Ergonomics, and the Design of Work* (3rd ed.), Taylor and Francis, London.

TABLE 5.2
Data collected on maximum grip strength (N) with the wrist at various joint angles

Flexion		Extension		Ulnar deviation		Radial deviation		Participant information
Angle	Force	Angle	Force	Angle	Force	Angle	Force	
%		%		%		%		Number
%		%		%		%		Mass
%		%		%		%		
%		%		%		%		Stature
%		%		%		%		
%		%		%		%		Number
%		%		%		%		Mass
%		%		%		%		
%		%		%		%		Stature
%		%		%		%		
%		%		%		%		Number
%		%		%		%		Mass
%		%		%		%		
%		%		%		%		Stature
%		%		%		%		

Van Cott, H. P. and Kincade, R.G., 1972, Human Engineering Guide to Equipment Design (Rev. ed.), Superintendent of Documents, U.S. Government Printing Office, Washington, D.C., 20402.

5.2 STATIC MUSCLE CONTRACTIONS

OBJECTIVES

- To see if the equations of Monod and Scherer, and Rohmert can be applied to box holding.
- To compare their suitability for this task and the relationship of body size to grip strength.
- To demonstrate some problems of static load.

APPARATUS

Two sets of five briefcase boxes (see Web site) of weights 100, 150, 200, 250, and 300 N
Grip strength dynamometer
Anthropometer
Balance-type scale
Stopwatches

TECHNICAL BACKGROUND

Maintaining a static muscle contraction is complicated by the fact that the contraction occludes the flow of blood to the muscles, thereby preventing fresh supply of oxygenated blood and the removal of lactic acid (Astrand and Rodahl 1986; Corlett 2005). Grandjean and Kroemer (1997) suggest time limits for the ability to maintain a static contraction for a number of different time intervals. Although limits can be appreciated conceptually in the classroom, it is another thing altogether to experience them for real. This experiment achieves that. Because of the extended holding times, students tend to remember this experience really well, and so have a much better appreciation of the desirability of tackling the problem.

The assumption is that holding time is proportional to grip strength, which appears to have fair validity, and the intention is to assess this in relation to the holding time curves derived by Monod and Scherer (1965) and Rohmert (1973). It is also assumed that grip strength MVC is the same as box-holding MVC, which is not strictly true as the grips and muscles involved are slightly different. At one time it was thought that there was an asymptote at 15% of MVC so that below this percentage a constant static contraction could be maintained more or less indefinitely. More recent thinking is that if there is any such limit, it must be very close to 0%.

Rohmert equation: $T = -90 + 126/p - 36/p^2 + 6/p^3$

Monod and Scherer equation: $T = 2.5/[(p - 0.14)^{2.4}]$

where p = proportion of MVC, and T = maximum endurance time.

Boxes are allocated to individuals to represent approximately 30, 50, and 70% of the grip strength MVC of each participant to get representative points on the curve. These are compared with the figures predicted by the equations of Monod and Scherer, and Rohmert. By suitable sorting of the data per Table 5.3, the students can see how to extract further meaning from their results.

TABLE 5.3
Mean times (cmin) for lifts at various decades of MVC

Decade of %MVC	First lift	Second lift	Third lift	Means for decades
20–29				
30–39				
40–49				
50–59				
60–69				
70–79				
Means for lifts				

PROCEDURE

1. Measure the stature and mass of all members of the class, and record them in Table 5.4.
2. Zero the dynamometer, and then take two readings of the grip strength of the dominant hand of each person, at least 5 minutes apart. Record your results in Table 5.4.

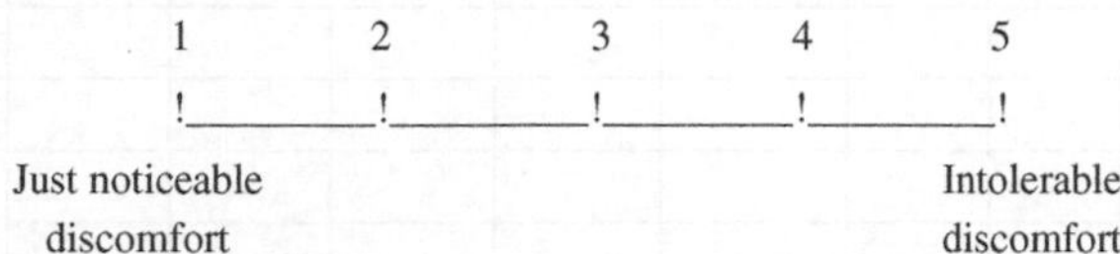

3. Select three boxes that represent low, middle, and high percentages of the grip strength MVC of each participant. Each participant then holds the first box of their sequence in the dominant hand until they reach 5 according to the above discomfort scale (say, as though carrying it into town). Record the time taken in Table 5.4.
4. After at least 10 minutes of rest, proceed with the next load in the sequence in as much balanced an order as possible across participants to minimise order and fatigue effects.

REQUIREMENTS

1. A short laboratory report.
2. Group data to draw up your version of Table 5.3.
3. Draw up the data in a table similar to Table 5.4, except with %MVC grip strength instead of box number.
4. Plot the holding time against %MVC on the top half of A4 graph paper or by computer, and fit a curve through the points (approximately). Use different symbols for the plots of the first, second, and third lifts. Then, plot two curves on this for values of 0.2, 0.3, 0.5, 0.7, and 0.9 for p in the equations of Rohmert, and Monod and Scherer. Finally, plot the row means against mean %MVC from Table 5.3.
5. Examine the data to see if the order effect is eliminated, whether Rohmert's curve or Monod and Scherrer's curve fits the data in some way, or there is a difference between them, and whether grip strength was really the limiting factor.
6. Establish what relationships exist between grip strength and stature, and grip strength and mass.
7. Plot log(holding time) against %MVC on the lower part of the graph paper, and get the regression equation and correlation coefficient (and significances) for these.

TABLE 5.4
Grip strengths, boxes held, and holding times (cmin)

Participant			Grip test (N)		First hold		Second hold		Third hold	
Number	mm	kg	First	Second	Box	Time	Box	Time	Box	Time

REFERENCES

Astrand, P.O. and Rodahl, K., 1986, *Textbook of Work Physiology*, McGraw-Hill, New York.

Corlett, E.N., 2005, Static muscle loading and the evaluation of posture, In *Evaluation of Human Work* (3rd ed.), Wilson, J.R. and Corlett, E.N. (Eds.), Taylor and Francis, London, pp. 453–496.

Grandjean, E. and Kroemer, K.B.E., 1997, *Fitting the Task to the Human* (5th ed.), Taylor and Francis, London.

Monod, H. and Scherer, J., 1965, The work capacity of a synergic muscle group, *Ergonomics*, 8, 329–338.

Rohmert, W., 1973, Problems in determining rest allowances. 1. Use of modern methods to evaluate stress and strain in static work, *Applied Ergonomics*, 4, 91–95.

5.3 ANTHROPOMETRY INVESTIGATION

OBJECTIVES

- To see the difficulty of getting accurate and repeatable values
- To examine the usefulness of estimates from stature
- To compare class data with that from other populations

APPARATUS

Anthropometer (e.g., Holtain)
Balance scale
10-mm-thick boards or something similar to bring thighs approximately horizontal
Measuring seat with a flat sitting surface (see Web site for a drawing of an example)

TECHNICAL BACKGROUND

One of the fundamental concerns of ergonomics is to accommodate individual differences, and one of the most obvious and least controversial of these is that of body dimensions or anthropometry. It has obvious relevance to the design of tools, equipment, workplaces, furniture, vehicles, and so on. Exposure to it is also a good way to demonstrate some common ergonomics issues such as the problems of repeatability of measurements, numerical modelling, and statistical estimation.

Some texts imply that body dimensions can be predicted to an acceptable degree of accuracy from stature measurement. For that reason, a number of different dimensions are measured here to demonstrate the weakness of this procedure, particularly for lateral dimensions but less so for those in the vertical direction. Including the measurement of body mass provides a good opportunity to establish a small correlation matrix and, hence, raises issues related to such statistical matters. The dimensions have been chosen deliberately to use easily locatable body points and a simple anthropometer. For some reason, it always engenders a fair degree of enjoyment.

As an aside, the exercise helps to show how population dimensions have grown over time and how they differ between ethnic groups (e.g., Chapanis 1974). One of the interesting aspects of the latter is how the ratio of sitting height to stature differs ethnically, with important implications for the design of products for world markets, for example, automobile seating. Another feature that can be highlighted is the greater ratio of buttock width to stature among females compared to males.

PROCEDURE

1. Measure all class members for Mass and Stature, and record them as in Table 5.5.
2. Each participant then sits on the seat in the "Erect" position (i.e., not slumped), and boards are placed under the feet so that the thighs lie horizontally.
3. Then, measure these dimensions, and record them as in Table 5.5:
   - Sitting Height Erect (from seat surface to top of head) (*Note:* Ensure that the anthropometer is vertical)
   - Buttock Breadth Seated (*Note:* Keep knees together and ensure that pockets are empty)
   - Popliteal Height
   - Thigh Clearance Height
   - Buttock–knee length
4. Repeat the measurements on at least one participant to examine their repeatability.

REQUIREMENTS

1. A short laboratory report.
2. Provide a table of data collected.
3. Plot sitting height erect, buttock breadth seated, and body mass against stature, with the first two on the left-hand vertical axis from 200 to 1000 mm, and mass on the right-hand vertical axis from 20 to 120 kg or so.
4. Calculate least-squares regression lines for mass, sitting height erect, and buttock breadth seated against stature (use a computer package or the following equations), draw or plot these lines, and write above each line the regression equation for it.
5. Calculate the correlation coefficient for each with stature and body mass, and get their statistical significance. Use these to draw up a correlation matrix.
6. From standard deviations and assumptions of students' "*t*" distribution, estimate the 5th and 95th percentiles for stature, mass, and buttock breadth seated. Compare these with the values from Konz and Johnson (2008), the U.S. Civilians data in Sanders and McCormick (1992), and civilian data in Pheasant and Haslegrave (2006). Present your comparisons in a table.
7. Use the *t*-test to compare the data means with each mean of the same data sets.
8. Comment on the suitability of stature as a predictor for these variables, and compare sitting height erect as a proportion of stature for yourselves with other ethnic groups.

TABLE 5.5
Anthropometry data collected

Participant number	Stature (mm)	Body mass (kg)	Sitting height erect (mm)	Popliteal height (mm)	Thigh clear-ance Ht (mm)	Buttock–knee length (mm)	Buttock breadth seated (mm)	Sitting height–stature ratio

9. Get regression equations for popliteal height, thigh clearance height, and buttock–knee length in relation to stature. How good are they?

REFERENCES

Chapanis, A., 1974, National and cultural variables in ergonomics, *Ergonomics*, 17, 153–175.

Konz, S. and Johnson, S.L., 2008, *Work Design: Occupational Ergonomics* (6th ed.), Holcomb Hathaway, Scottsdale, AZ.

Pheasant, S. and Haslegrave, C.M., 2006, *Bodyspace: Anthropometry, Ergonomics, and the Design of Work* (3rd ed.), Taylor and Francis, London.

Sanders, M.S. and McCormick, E.J., 1992, *Human Factors in Engineering and Design* (7th ed.), McGraw-Hill, New York.

Regression line: $Y = a + b.X$

(*Note:* Y estimated from X, NOT the reverse.)

where $a = \dfrac{(\Sigma y)(\Sigma x^2) - (\Sigma x)(\Sigma xy)}{N.\Sigma x2 - (\Sigma x)^2}$ $b = \dfrac{N.\Sigma xy - (\Sigma x)(\Sigma y)}{N.\Sigma x2 - (\Sigma x)2}$

$$r = \frac{N.\Sigma xy - (\Sigma x)(\Sigma y)}{\sqrt{[N.\Sigma x^2 - (\Sigma x)^2].[N.\Sigma y^2 - (\Sigma y)^2]}}$$

Significance of the Correlation Coefficient

Look up values in Murdoch and Barnes (for example). For total correlation, the degrees of freedom amount to 2 less than the number of data pairs. The probabilities at the head of the columns refer to the two-tail test of significance, and give the probability that *r* will be greater than the tabulated values purely due to chance effects. For a single-tail test, the probabilities should be halved, as usual.

If the chance probability is greater than 0.05, we usually decide that the result is not significant; that is, it could well have arisen by chance, so we decide that it is not a significant correlation.

5.4 POSTURE ANALYSIS

OBJECTIVES

- To learn to use RULA and Drury's technique
- To analyse the posture of a typical industrial job
- To devise an improved posture

- To devise ways to reduce static load problems and cumulative trauma disorders
- To estimate the improvements from improved postures

APPARATUS

Video recordings of a variety of industrial jobs or of laboratory tasks
Video players and monitors
Manual goniometers
RULA recording and scoring charts

TECHNICAL BACKGROUND

Most cumulative trauma disorders (or repetitive strain injuries) result from one or more of the following: the force applied, the frequency of application, postural deficiencies, and lack of recovery time (Putz-Anderson 1988). These cause repeated microtraumas of the soft tissues. Postural deficiencies tend to result from awkward joint angles that are somewhat removed from neutral (Corlett 2005). These problems tend to be acute in the spine because the intervertebral discs have no blood supply, and hence, no direct means to provide nutrients or remove waste products. For these functions, they require changes in intervertebral pressure, and therefore, regular changes in posture. Design work in ergonomics has to find ways to position the seat/bench system; design tools and positions of tools and/or components used in the occupation so that the joint angles are reduced to acceptable levels, preferably neutral; and to induce regular changes in posture. Similarly, it is also necessary to reduce the magnitudes of the forces exerted and their rate of repetition, but these effects can be compensated to some extent by providing rest and recovery opportunities.

The simple RULA technique was devised by McAtamney and Corlett (1993), and for survey work on a complete factory or shop, it is very suitable. It provides pictorial representations of ranges of joint angles and a composite scoring system to assess acceptability and/or corrective actions required. However, where greater detail is required, something like Drury's (1987) method is needed. He divided Ranges of Motion (ROM) into zone 0, with negligible exposure to risk (from neutral to ±10% of ROM), zone 1, with low risk (±10% to ±25% of ROM), zone 2, with moderate risk (±25% to ±50% of ROM), and zone 3, with severe risk (> ±50% of ROM). Average ROM values are taken from NASA data. The approach here is to use both methods in combination. For a greater range of such tools, students should consult the MIRTH software package (see Web site).

The aim in this exercise is to gain experience in observing conditions of this type, and to think up ways in which ameliorative measures can be implemented (Bergamasco et al. 1998). Unfortunately, the forces cannot be measured when just observing a video; but in some instances, it may be possible to deduce ways to reduce masses held or moved. The exercise also familiarises students with much of the anatomical terminology in the area and the limitations on human movements in some body regions, especially the wrist.

A full investigation needs to involve multiple views of the worker at work, but time constraints dictate that these are not practical in a laboratory exercise.

However, the technique for one view or several views is the same, so the learning obtained remains valid. If time permits, time data can also be extracted to get a picture of the daily tasks performed by the operator for use in a time-study-type analysis, or simply to calculate Drury's Daily Damaging Wrist Motions (DDWM), that is, the total number of times nonzero wrist exposures would occur in a notional 8 hour shift.

PROCEDURE

1. Split into groups of three.
2. View the video several times to gain familiarity with the job.
3. Devise a breakdown of the job into three groupings of relevant motion elements.
4. Analyse these three elements using RULA, get the scores, and decide which one is worst. Then proceed to the Drury analysis.
5. For the worst element, determine the extreme joint angle for each joint in two or three subtasks. Set the arms of the manual goniometer to match the angle made by the joint in question. Those in the plane of the screen can be measured on the screen, but for the others, the goniometer must be held in some other plane to simulate that of the joint elements, and viewed from the same angle as that of the camera. Read off the angle from the built-in protractor.
6. Record the value found for each subtask on the Drury Posture Description sheet (Table 5.6), and use a second copy of the sheet to record the class of the joint angle values in joint zone scores (Table 5.8).
7. Examine all the elements for damaging wrist motions, and get their frequency.
8. Determine the job cycle time and the number of times per cycle that each element occurs.
9. Note the general aspects of postures and grips to complete Table 5.7.
10. Examine the job in detail for likely problems due to static load and Cumulative Trauma Disorders (CTDs). You may need to use your imagination here.

REQUIREMENTS

1. A long laboratory report describing the job studied.
2. Provide a table giving RULA and Drury element descriptions with the frequency per cycle for each, the time duration of each, and the RULA scores for the relevant elements (including actions advised).
3. On the same table, give data values and calculations for DDWMs for left and right hands, assuming that the job lasts for 8 hours with no breaks; for comparison, do it also (on the same table) for your proposed improved workplace design.
4. Provide another table listing the joints concerned and their postures with their classification zones and the grips used.
5. Submit a table, listing in the LH column the postural deficiencies you have detected in the work design, and in the RH column, your design changes to remedy them.

TABLE 5.6
Posture description/analysis sheet 1
Job title:

Subtask no.		**1**	**2**	**3**	**4**	**5**	**6**	**7**	**8**	**9**	**10**
LABEL											
FREQUENCY/group											
NECK: rotation at joint											
Lateral bend (to side)	L/R										
Flexion/extension (flexion =											
fwd, extension = bwd)											
BACK: rotation of torso											
Lateral bend (to side)	L/R										
Flexion/extension (flexion =											
fwd, extension = bwd)											
SHOULDER JOINT											
Arms: adduction/abduction	L										
(abduction = arm out from V)	R										
Upper arm rotation—out/in	L										
(in = thumb towards body)	R										
Flexion/extension (flexion =	L										
upper arm forward from V)	R										
ELBOW: flexion	L										
(flexion = decrease angle)	R										
FOREARM: pronation/	L										
supination (pronation =	R										
rotate hand inwards)											
WRIST: flexion/extension	L										
(flexion = bend hand down)	R										
Deviation—radial/ulnar	L										
(radial = bend toward thumb)	R										
LEGS: thigh to horiz	L										
thigh to horiz	R										
Shin: to vertical	L										
to vertical	R										
Foot: to horizontal	L										
to horizontal	R										
Rotation to the side	L										
Rotation to the side	R										

Note: fwd = forwards, bwd = backwards, V = Vertical, Horiz = Horizontal.
Mark each angle with a letter for the posture, for example, U for Ulnar deviation.
Source: Adapted with permission from Drury, C.G., 1987, A biomechanical evaluation of the repetitive motion injury potential of industrial jobs, *Seminars in Occupational Medicine*, 2, March, 41–49, Thieme Medical Publishers, New York.

TABLE 5.7
Posture description/analysis sheet 2

Job title:

Subtask no.		1	2	3	4	5	6	7	8	9	10
LABEL											
FREQUENCY/group											
POSTURE sit/stand											
Armrest (Y/N)											
Foot pedal (Y/N)											
Backrest (Y/N)											
GRIPS Power	L										
	R										
Precision	L										
	R										
Pulp pinch	L										
	R										
Lateral pinch	L										
	R										
Tip	L										
	R										
Other	L										
	R										
HAND/ push/pull	L										
ARM	R										
FORCES up/down	L										
	R										
in/out	L										
(to body CL)	R										
FOOT/LEG	L										
FORCES	R										
VIBRATION	L										
	R										
SHOCK	L										
	R										

Source: Adapted with permission from Drury, C.G., 1987, A biomechanical evaluation of the repetitive motion injury potential of industrial jobs, *Seminars in Occupational Medicine*, 2, March, 41–49, Thieme Medical Publishers, New York.

TABLE 5.8
Posture classification data

	Zones for joint angles (degrees)			
	0	**1**	**2**	**3**
NECK				
Rotation of joint	0–8	8–20	20–40	40+
Lateral bend (away to side)	0–5	5–12	12–24	24+
Flexion (towards front)	0–6	6–15	15–30	30+
Extension (towards back)	0–9	9–22	22–45	45+
BACK				
Rotation of upper torso	0–10	10–25	25–45	45+
Lateral bend (away to side)	0–5	5–10	10–20	20+
Flexion (towards front)	0–10	10–25	25–45	45+
Extension (towards back)	0–5	5–10	10–20	20+
SHOULDER JOINT				
Arms—Abduction (arm away from CL)	0–13	13–34	34–67	67+
Arms—Adduction (arm towards CL)	0–5	5–12	12–24	24+
Joint—Rotation—out (thumb out from CL)	0–3	3–9	9–17	17+
Joint—Rotation—in (thumb towards CL)	0–10	10–24	24–49	49+
Arm—Flexion (upper arm fwd of V)	0–19	19–47	47–94	94+
Arm—Extension (upper arm bwd of V)	0–6	6–15	15–31	31+
ELBOW				
Flexion	0–14	14–36	36–71	71+
FOREARM				
Pronation (rotate hand inwards)	0–8	8–19	19–39	39+
Supination (rotate hand outwards)	0–11	11–28	28–57	57+
WRIST				
Flexion (bend hand down)	0–9	9–23	23–45	45+
Extension (bend hand up)	0–10	10–25	25–50	50+
Deviation—Radial (towards thumb)	0–3	3–7	7–14	14+
Deviation—Ulnar (away from thumb)	0–5	5–12	12–24	24+

Notes: V = Vertical, CL = Centre Line, fwd = forwards, bwd = backwards.

For a further explanation of the terms see Appendix A of Vern Putz-Anderson (Ed.), 1988, *Cumulative Trauma Disorders,* Taylor and Francis, London, 115–117.

Source: Devised by Drury, C.G., 1987, in A biomechanical evaluation of the repetitive motion injury potential of industrial jobs, *Seminars in Occupational Medicine*, 2, March, 41–49, Thieme Medical Publishers, New York, from data in NASA, 1978, Anthropometric Source Book, NASA Reference Publication 1024, authored by L.L. Laubach; adapted with permission.

6. Compile a table of static load problems with (alongside) suggested remedies for each.
7. Examine the motions involved for causes of CTDs, and devise ways to reduce or eliminate them. List the problems you have found and your countermeasures in a separate table.
8. Submit a drawing of the workplace design as seen on the video (top half of the page) and your new improved one (bottom half of the page).
9. Complete a Posture Description/Analysis sheet for the task as seen, estimate the angles and zones that will result from your new layout, and record them on a fresh Posture Description/Analysis sheet, with the expected zones for the joints.
10. Further, in addition to the usual points, consider the suitability of the techniques, such as your chances of getting the same action categories (for RULA) and the same angle and zone values (for Drury) if you did it all over again (repeatability), and how different these might be if the person studied were to be taller or shorter (i.e., generality of results).

REFERENCES

Bergamasco, R., Girola, C., and Colombini, D., 1998, Guidelines for designing jobs featuring repetitive tasks, *Ergonomics*, 41, 1364–1383 (Special Issue on Occupational Musculoskeletal Disorders).

Corlett, E.N., 2005, Static muscle loading and the evaluation of posture, In *Evaluation of Human Work* (3rd ed.), J.R. Wilson and E.N. Corlett (Eds.), CRC Press, Boca Raton, FL, 453–496.

Drury, C.G., 1987, A biomechanical evaluation of the repetitive motion injury potential of industrial jobs, *Seminars in Occupational Medicine*, 2, 1, 41–49 (Publisher: Thieme Medical Publishers, New York.)

McAtamney, L. and Corlett, E.N., 1993, RULA: a survey method for the investigation of work-related upper limb disorders, *Applied Ergonomics*, 24, 91–99.

Putz-Anderson, V., 1988, *Cumulative Trauma Disorders*, Taylor and Francis, London.

www.rula.co.uk for info on RULA. Download the off-line version form for use here.

http://www.ergonomics.ie/mirth.html provides access to aids developed under the EU-funded MIRTH (Musculoskeletal Injury Reduction Tool for Health and safety) project.

5.5 MAXIMUM OXYGEN UPTAKE

OBJECTIVES

- To gain experience in using specialised lab equipment
- To learn how to measure the maximal oxygen uptake of people
- To become alerted to the difficulties and errors in such measurement

APPARATUS

Bicycle ergometer (e.g., Monark)
Polar Tester Heart Rate Monitor
Anthropometer
Cardiovascular O_2/CO_2 analyser
Mouthpieces and pneumotachs
Instrument to measure humidity and temperatures
Barometer
Balance beam scale
RPE scale
Spirometer (e.g., Vitalometer)
Steriliser beaker and fluid

TECHNICAL BACKGROUND

Tasks that are physically demanding have been eliminated or considerably reduced in most industrialised countries. Nevertheless, less demanding physical tasks still exist and the fatiguing effects depend partly on the ability of the person to get oxygen from the air into the lungs and then to the muscles in question, and then to reoxygenate their blood (Astrand and Rodahl, 1986). The magnitude of the effect depends on the relationship between the job demands and the physiological capacity of the person, which means that the latter (maximum oxygen uptake, VO_{2max}) needs to be determined.

To keep risks low, the preferred approach is one of submaximal testing, as described by Sinning (1975). In the process, the students learn the precautions and procedures required. The results provide an opportunity to examine some relationships between human physical measures and VO_{2max} and at least one method for estimating it from simpler measures. Although the small sample provides an inadequate base for good predictions, the results tend to show some aspects of the scatter likely in such data, and hence, the need for caution in using them.

PROCEDURE

1. Choose four participants (preferably of different sizes) to complete a health questionnaire and informed consent form, and then measure the vital capacity of each using the Spirometer. If the results are acceptable, proceed with the experiment.
2. Turn on the analyser gas bottles and the pump, and calibrate the analyser. Get T_{db}, T_{nwb}, and barometric pressure, and then measure the body mass and stature of the four class members. Record the values in Table 5.9.
3. Wet the electrodes area of the Polar Tester, and then strap it around the first participant just below the pectoral muscles. Start the Polar watch. Wait for HR to settle at a steady value, and then record it as the resting value in Table 5.9.
4. Fit the nose clip and mouthpiece, and connect the latter to the sampling line. Enter the participant's data into the computer, and wait for things to settle down.
5. Set the ergometer load to demand 50 W, and then have the participant pedal for at least 6 minutes. During the last 2 minutes, note heart rate and RPE and save the values to be printed out afterwards for completion of Table 5.9.
6. Repeat with the remaining participants.

TABLE 5.9
Data collected at the lower level of power

Power = 50 W	Participant 1	Participant 2	Participant 3	Participant 4
Stature (mm)				
Mass (kg)				
HR_{rest} (b.p.m.)				
HR_{work} (b.p.m.)				
RQ				
VO_2 (L/min)				
T_{nwb} (C)				
T_{db} (C)				

Barometer reading =

TABLE 5.10
Data from the test at higher power

Power = 100 W	Participant 1	Participant 2	Participant 3	Participant 4
HR_{work} (b.p.m.)				
RQ				
VO_2 (L/min)				
T_{nwb} (C)				
T_{db} (C)				

7. Then raise the demand to 100 W and repeat it on all the participants, recording the data in Table 5.10.
8. After each participant, detach the gas lines and the sampling lines and use the former to blow down the latter and sterilise the mouthpieces, flowmeter piece, and nose clip.

REQUIREMENTS

1. A long laboratory report.
2. Estimate VO_{2max} from the two sets of data on each participant, at STPD (Standard Temperature and Pressure Dry).
3. Use Konz and Johnson's (2008) equations to estimate the same values and compare them.
4. Carry out a regression analysis of VO_2 against heart rate across all participants, establish the correlation coefficient and its significance, and plot a graph of the results for all participants.
5. Use the specific value of VO_{2max} to classify the fitness of the participants; compare these to the Konz and Johnson values.
6. Calculate the regression line and correlation between VO_{2max} and vital capacity, stature, and mass. How well do the latter predict the value?

REFERENCES

Astrand, P.-O. and Rodahl, K., 1986, *Textbook of Work Physiology* (3rd ed.), McGraw-Hill, New York.

Konz, S. and Johnson, S.L., 2008, *Work Design: Occupational Ergonomics* (6th ed.), Holcomb Hathaway, Scottsdale, AZ.

Sinning, W. E., 1975, *Experiments and Demonstrations in Exercise Physiology*, Saunders, Philadelphia.

5.6 ENERGY EXPENDITURE

OBJECTIVES

- To examine the suitability for local students of the Pimental and Pandolf equations
- To examine the suitability of Konz's equations
- To gain experience in using Borg's Rating of Perceived Exertion (RPE; Borg 1982)
- To compare the quality of the estimates achieved

APPARATUS

Treadmill (e.g., Powerjog)	Barometer
Whirling psychrometer and RPE scale	Polar Tester heart rate monitor
Anthropometer	Spirometer (e.g., Vitalograph)
Scale (balance beam type)	Cardiovascular O_2/CO_2 analyser
Mouthpieces, pneumotachs, steriliser beaker	Steriliser fluid and bath (e.g., Milton)
Instrument to measure humidity and temperature (e.g., SCANTEC WIBGeT)	

A briefcase box (see Web site) of 10 kg mass

TECHNICAL BACKGROUND

In everyday manual handling tasks, workers often walk with a load and it is necessary to assess the effects of these tasks to see if the effort falls within or outside safe limits. Such calculations can also provide a means to estimate recovery requirements once the task has been completed. The predictor equation of Pimental and Pandolf (1979) is as follows, with slightly altered notation:

$$M = 1.5\ MS + 2.0(MS + L)(L/MS)^2 + \eta(MS + L)(1.5\ V^2 + 0.35 * V.G)$$

where the notation has been changed from the original to the following terms:

M = total metabolic cost (watts)
MS = mass of participant (kg)
L = load carried by the participant (kg)

η = terrain coefficient with the following values:

1.0 for a treadmill
1.2 for a hard surfaced road
1.5 for a ploughed field
1.6 for hard snow
1.8 for sand dunes

These were developed on young military recruits at the end of their first 3 months of basic training, so they were very fit and the equations may not be suitable for many worker populations.

The aim is to examine predictions from these equations, and those of Konz and Johnson (2008), against actual true readings of energy expenditure using some members of the class. The laboratory environment will be similar to that which obtains in many work situations but will be at the lower end of the scale in terms of environmental demands on the participants.

PROCEDURE

1. Get two or more participants to complete a health questionnaire and informed consent form, and then measure the vital capacity of each using the spirometer. Do not proceed with the experiment unless the results are acceptable.
2. Turn on the gas bottles and the pump, and calibrate the analyser. Measure the barometric pressure, and the wet and dry bulb temperatures. Then measure the body mass and stature of the participants, and enter them in Table 5.11.

TABLE 5.11
Results from walking with a load on a level surface

G = 0	Participant 1	Participant 2	Participant 3	Participant 4
Stature (mm)				
Mass (kg)				
HR_{rest} (b.p.m.)				
HR_{work} (b.p.m.)				
RQ				
VO_2 (L/min)				
T_{nwb} (C)				
T_{db} (C)				
RPE vote				

Barometer reading =

3. Wet the electrodes area of the Polar Tester, strap it around the participant just below the pectoral muscles, and start the tester. After waiting for the heart rate to settle at a steady value, record it as the resting value in Table 5.11.
4. Fit the nose clip and mouthpiece, and connect the latter to the sampling line. Enter the climate and participant data to the computer where relevant. Wait for things to settle down. Set the slope at G = 0, and then walk the participant on the treadmill at 4.8 km/h (see Konz and Johnson ACTFMT). Hand over a box of 10 kg mass, and walk for at least 5 minutes. During the last 2 minutes, the participant votes on the RPE, the steady HR is noted, and values are saved (Table 5.11).
5. Then detach the gas lines and the sampling lines and blow down the latter with the former, sterilise the mouthpieces, etc.
6. Repeat with the remaining participants.
7. Raise the slope to G = 10, and repeat it all. Record these data in Table 5.12.

REQUIREMENTS

1. A long laboratory report.
2. Use the Pimental and Pandolf equation to estimate M, and usc Konz and Johnson equations for BASLMT, ACTMET, and SDAMET, and so TOTMET.
3. Estimate INCHR from Konz and Johnson, and compare it to actual HR increase.
4. Estimate OXUPTK and PERFAT from the Konz and Johnson equations, and compare to actual data to check on fitness of participant—for interest.
5. Establish Brouha's rating of this work, and compare actual HR with value obtained from the RPE score.
6. Plot a graph for all participants, showing actual HR increase versus INCHR value from Konz and Johnson. How close are they to a line at 45°?
7. Calculate the actual metabolic rates from the RQ and VO_2 readings (at STPD), and compare these to the estimated values.

TABLE 5.12
Results from walking with a load on a slope

G = 10	Participant 1	Participant 2	Participant 3	Participant 4
HR_{work} (b.p.m.)				
RQ				
VO_2 (L/min)				
T_{nwb} (C)				
T_{db} (C)				
RPE vote				
Brouha's ratings				

TABLE 5.13
Data for example on Pimental and Pandolf equation

Who	Stature (mm)	Mass (kg)	HR_{rest} (bpm)	HR_{work} (bpm)	RQ	VO_2 (l/min)	T_{wb} (C)	T_{db} (C)	RPE	HR_{rec} (bpm)
25 yr	1776	78.5	60	112	0.7	1.7	15.1	18.7	11	68
30 yr	1791	64.0	55	101	0.7	1.24	15.0	18.7	11	52

REFERENCES

Borg, G.A.V., 1982, Psychophysical bases of perceived exertion, *Medicine and Science in Sports and Exercise*, 14, 377–381.

Konz, S. and Johnson, S.L., 2008, *Work Design: Occupational Ergonomics* (6th ed.), Holcomb Hathaway, Scottsdale, AZ.

Pimental, N.A. and Pandolf, K.B., 1979, Energy expenditure while standing or walking slowly uphill or downhill with a load, *Ergonomics*, 22, 963–973.

TUTORIAL ON ENERGY EXPENDITURE

1. A 30-year-old lumberjack, with a body mass of 80 kg and stature 1800 mm, walks for 1 km at 3 km per hour up a 6% grade in a forest clearing to reach his worksite. The terrain coefficient is about 1.35. In the sun the dry bulb temperature is 32C, air velocity is 5 km/h directly into his face, and relative humidity is 70%. In his hands he carries an axe of mass 6 kg and on his back a pack of food and spare clothes of mass 8 kg. If external work is at a rate of 8 W, use appropriate techniques to determine the gross metabolic cost of this exertion.
2. A 40-year-old male walks on a 10% slope at 4.8 km/h carrying a 10 kg box for 5 minutes, and at the end his heart rate reached 137 beats/min. His resting HR was 65, his stature 1830 mm, and body mass 89 kg. His oxygen consumption was 2.48 L/min with an RQ of 0.7, the wet bulb temperature was 15.1C, and dry bulb was 18.9C, and he voted it as 13 on Borg's RPE scale. After 5 minutes of recovery, his HR was 79. Use Pimental and Pandolf equations to estimate M, and compare it to the value measured from O_2 consumption.
3. Two males aged 25 and 30 years did the preceding task but on the flat, and gave the results in Table 5.13. Use Pimental and Pandolf equations to estimate M, and compare it to the value determined from O_2 consumption.
4. A 20-year-old female with "good" fitness and body mass of 60 kg walks with a light load. Her stature is 1650 mm. If she has to do this for an 8 hour shift, what would her HR be at the end according to the Konz and Johnson equations, if her HR_{rest} is 80 beats per minute?
5. A 45-year-old male of 80 kg body mass and 1750 mm stature carries a 20 kg suitcase at 5 km/h on a hard surface up a 10% slope. Assuming a basal metabolic rate of 100 W, what is his estimated TOTMET from the Konz and Johnson equation using Table of Energy cost for various activities and

making allowances for slope and load, and what is the estimate of M from the Pimental and Pandolf equation, assuming he does it for long enough to reach equilibrium?

6. A 55-year-old male of poor fitness works at a job rated as hard. He has a stature of 1812 mm and body mass of 90 kg. What is the estimated total metabolic cost for him in doing this job, his estimated final HR if resting HR level is 80 b.p.m., and his estimated percentage body fat? What is Brouha's rating for his work? Make suitable estimates of ENERGY value and percentage of VO_{2max} used for the OXUPTK value in the equation.

5.7 HEAT DISSIPATION

OBJECTIVES

- To examine how well the Givoni and Goldman equations predict the effects of heat on students
- To compare the effects of a high-humidity environment to one of radiant heat
- To compare some measures of the thermal environment

APPARATUS

Hot-box (see Web site for an example) with container of boiling water	Hot wire anemometer for air velocity
	Polar Tester Heart Rate Monitor (two)
Two step-test boxes (see Web site)*	Batteries for the bicycle
Bicycle ergometer	Sling psychrometer
Radiant heaters (4 off 1 kW)	Anthropometer
Stopwatches	Metronome
Botsball thermometer (for T_{bot})	Scantec WiBGeT (or similar)
Cardiovascular O_2/CO_2 analyser	Balance scale for body mass
Steriliser flask and towel	Spirometer for vital capacity

**Note*: P_{ex} = MS*9.81*20/60*(box height in meters), where MS = body mass (kg) approximately.

TECHNICAL BACKGROUND

Doing heavy physical work in a hot and humid environment has always been difficult for human workers, and the ability to cope with heat stress has long been sought among prospective employees, even though there is less demand for it today in industrialised countries. It is therefore important to be able to predict with some accuracy the likely stress levels and the recovery requirements. Similarly, with the advent of air conditioning, it is possible to estimate the benefits in recovery that can be obtained from that. These can be used to determine suitable work–rest schedules. Givoni and Goldman (1971, 1972, 1973) developed a set of equations for this work. Good sources for information on this subject are Parsons (2003) and Parsons (2005).

Body effort to produce useful work generates heat over and above that needed to support basal metabolic activity. For thermal equilibrium to be maintained, this extra heat has to be dissipated, which can be by evaporation, radiation, or convection. However, at the same time, the body may be gaining heat from the environment owing to radiation or convection. So, we have the heat balance equation, which is expressed as follows:

$$S = M - E \pm R \pm C - P_{ex} \text{ watts}$$

where S = total heat stored in the body
M = total metabolic cost measured from oxygen uptake
E = heat lost through evaporation from the body surface
R = heat gained or lost through radiation
C = heat gained or lost through convection
P_{ex} = external useful power produced by the body in doing the job

For example, when walking up a slope with a load *m*, we will have

$$P_{ex} = 0.0981.m.v.G$$

with m = mass moved (kg)
v = velocity (m/s)
G = grade of slope % (i.e., 20% = 20, not 0.20).

Note: In theory, heat could be gained or lost through conduction, but this is negligible except for an activity such as swimming, which is unlikely to be of interest here.

However, their participants were military trainees who had just completed 3 months of basic camp, they were young, and they were from the U.S.; therefore, they may have been ethnically different from workers in some other parts of the world. Also, the clothing ensembles tested were of a limited variety, although they did cover a good range. This experiment tests the suitability of these equations for the class members in two fairly difficult climates, one of high humidity and one of high radiant heat. As the environment cannot be properly controlled, it is to some extent a demonstration but very informative nevertheless.

PROCEDURE

Common Features

1. All participants complete a questionnaire on their health, and sign an informed consent form.
2. Measure participants' vital capacity, and get a report on their pulmonary health.
3. If their health is OK, measure the mass and stature of the participants and complete the details in Table 5.14. Moisten the Polar Tester surfaces, and attach it to the participant.
4. Rest the participants until the HR is steady, and record it (Table 5.15). Use the sling psychrometer (stationary) or instrument to get T_{nwb} and T_{db} temps. Also get T_g and T_{bot} and WBGT, and record them.

TABLE 5.14
Participant data for heat dissipation experiment

Group no.__________ Course__________ Date__________

Participant no.	1	2	3	4
Mass (kg)				
Stature (mm)				

TABLE 5.15
Thermal data from heat dissipation experiment the hot box

WHO	HR_{rest}	Con	HR_f (b.p.m.)	T_{nwb}	T_{db}	T_g	T_{bot}	RER (RQ)	VO_2 (L/min)	HR_{rec} (b.p.m.) 5 min
		Cool								
XXX	XX	Hot								
		Cool								
XXX	XX	Hot								

Note:
1. RER = Respiratory Exchange Ratio = ratio of VCO_2 to VO_2 when these are measured at the mouth. In steady-state conditions, it is approximately the same as RQ.
2. RQ = Respiratory Quotient = ratio of VCO_2 to VO_2 during resting steady state.

High Humidity (Hot-Box) Test Specifics

5. Sterilise the bite-piece and fit it, and then wait for the analyser to settle down. With the metronome set at 40/min S steps onto the box (20 times/min) and off it (20 times/min) in the lab at this rate for 5 minutes.
6. At the end, note HR_{max}, and get CO_2 and VO_2 and flowrate values from the printer of the analyser (if fitted) or in Table 5.15.
7. Leave the participant for 5 minutes to recover and then note HR_{rec} (Table 5.15).
8. Then repeat all the earlier steps in the hot-box.

Radiant Heat (Bicycle) Test Specifics

9. After obtaining the resting HR for the second participant, sterilise the bite-piece, dry it, and fit it. Then he or she rides the bicycle at 100 W for 5 minutes.
10. At the end, note HR_{max}, CO_2, and VO_2, and flowrate (Table 5.16). Record T_{nwb}, T_{db}, T_g, T_{bot}, and WBGT.
11. Leave the participant for 5 minutes to recover, and then note HR_{rec} (Table 5.16).
12. Once HR_{rest} has been reached again, repeat all of this with the radiant heaters switched on.

Note: Position the heaters at 0.5 m from the participant's torso.
Repeat tests: Use more participants as time permits.

TABLE 5.16
Thermal data from heat dissipation experiment for the bicycle

WHO	HR_{rest}	CON	HR_f (b.p.m.)	T_{nwb} (C)	T_{db} (C)	T_g (C)	T_{bot} (C)	RER (RQ)	VO_2 (L/min)	HR_{rec} (b.p.m.) 5 min
S3		Cool								
XXX	XX	Hot								
S4		Cool								
XXX	XX	Hot								

REQUIREMENTS

1. A long laboratory report.
2. Get predicted HR values from Givoni and Goldman equations after 5 minutes of work, for each condition.
3. Plot a graph to compare actual HR with predicted values with a line that goes through the origin and at 45° slope, to represent exact agreement; ensure that the scales are the same on each of the axes.
4. Calculate the Heat Stress Index (HSI, preferably using formulae from Parsons 2003) and the Index of Thermal Stress (ITS) from Parsons; compare them with WBGT and T_{bot} between cool and hot for both tasks. Do they show similar trends, etc.?
5. Compare work HR after 5 min between both conditions, and compare to equation estimates.
6. Estimate resting HR (from Givoni and Goldman) for the lab conditions, and compare them to the actual HR_{rest}.
7. Discuss the effects of humid versus radiant environments.

REFERENCES

Givoni, B. and Goldman, R.F., 1971, Predicting metabolic energy cost, *Journal of Applied Physiology*, 30, 429–433.

Givoni, B. and Goldman, R.F., 1972, Predicting rectal temperature response to work, environment, and clothing, *Journal of Applied Physiology*, 32, 812–822.

Givoni, B. and Goldman, R.F., 1973, Predicting heart rate response to work, environment, and clothing, *Journal of Applied Physiology*, 34, 201–204.

Parsons, K.C., 2003, *Human Thermal Environments*, 2nd ed., Taylor and Francis, London.

Parsons, K., 2005, Ergonomics assessment of thermal environments, In *Evaluation of Human Work* (3rd ed.), Wilson, J.R. and Corlett, E.N. (Eds.), CRC Press, Boca Raton, FL.

TABLE 5.17
Data sheet for clothing values

Clothing type	Insulation value (clo)	Permeability index ÷ insulation value (im/clo)
Shorts only	$0.57(v_{eff})^{-0.30}$	$1.20(v_{eff})^{0.30}$
Shorts plus short-sleeved shirt	$0.74(v_{eff})^{-0.28}$	$0.94(v_{eff})^{0.28}$
Cotton trousers plus long-sleeved shirt	$0.99(v_{eff})^{-0.25}$	$0.75(v_{eff})^{0.25}$
Above plus heavier coverall	$1.50(v_{eff})^{-0.20}$	$0.51(v_{eff})^{0.20}$

$v_{eff} = v_{air} + 0.004(M - 105)$, *where M = total metabolic rate.*
Note: If walking into wind, v_{air} = walk speed + air speed (m/s).
1 clo = the clothing insulation required to keep a resting man indefinitely comfortable at 21C, RH < 50%, and air velocity = 0.1 m/s.
Source: From Givoni, B. and Goldman, R.F., 1972, *Journal of Applied Physiology*, 32, 812–822, used with permission.

Givoni and Goldman Equations

$M_{net} = M - P_{ex}$ = net metabolic rate that goes into body heat,

where P_{ex} = external power generated

$(R + C) = (11.6/clo)(Ta - 36)$,

where $(R + C)$ = heat lost or gained by radiation and convection

clo = insulation value for person's clothing (see Table 5.17)

T_a = ambient temperature = dry bulb temperature (C)

E_{req} = required rate of evaporative cooling = $M_{net} + (R + C)$

E_{cap} = capacity rate for evaporative cooling = $25.5\ (i_m/clo)(44 - VP_{amb})$

where VP_{amb} = vapour pressure ambient (mm Hg)

(for i_m/clo values, see Table 5.17)

$CP_{eff} = 0.27(i_m/clo)(44 - VP_{amb}) + (0.174/clo)(36 - T_a) - 1.57$

= cooling power effective of the environment

T_{re} final $= 36.75 + 0.004\ (M - P_{ex}) + 0.025/clo(T_a - 36) + 0.8.e^{\,0.0047(E_{req} - E_{cap})}$

= estimated final rectal temperature (as the best indicator of core body temp)

(for clo values see Table 5.17)

They also developed equations for estimating the heart rate after a period of work. The first step in these is to get an index of the heart rate (I_{HR}):

$I_{HR} = 0.4\ M + (2.5/clo)(T_a - 36) + 80.e^{\,0.0047(E_{req} - E_{cap})}$

= index of equilibrium heart rate level

From this, the final heart rate level (HR_f) is found, if the human continues indefinitely at this task, from one of two formulae as follows:

$$\text{If } 25 < I_{HR} < 225\text{: } HR_f = 65 + 0.35(I_{HR} - 25)$$

$$\text{and if } I_{HR} > 225\text{: } HR_f = 135 + 42[1 - e^{-(I_{HR} - 225)}]$$

Then, this value is used to find the heart rate after t minutes of work at the task, from

$$HR_{t(w)} = 65 + (HR_{f(w)} - 65)[1 - 0.8.e^{-(6 - 0.03(HR_{f(w)} - 65))t}]$$

where 65 = assumed resting heart rate in standard conditions (e.g., 20C and RH = 50%)

t = time from start of performing the task (hours)

$HR_{f(w)}$ = HR final at work.

TUTORIAL ON THERMAL ENVIRONMENT

1. A 50-year-old man does a manual materials-handling job at a rate of 130 W during which he breathes in oxygen at the rate of 2.0 litres per minute at an RQ of 0.8, and it takes him 5 minutes to do this job. He wears shorts and a short-sleeved shirt, the ambient temperature is 30C, ambient vapour pressure is 36 mm Hg, and the air velocity is 0.1 m/s. Using Givoni and Goldman's equations, determine:
   1. His heart rate at the end of the 5 minutes of work.
   2. His heart rate after 10 minutes of recovery in the same environment.

   Comment on his suitability for this job if he were to work at it for at least 2 hours.
2. For Problem 1 in the Energy Expenditure tutorial, calculate the Heat Stress Index if we assume that T_o = Dry Bulb Temp. *Note:* If we have high radiant heat, we should use Tg.
3. Taking the results of Problem 2 in the previous tutorial on Energy Expenditure, use Givoni and Goldman equations to estimate HR after 5 minutes of work and then 5 minutes of recovery, and compare them to the actual values. How does the estimate of HR from Borg's RPE compare with the actual measured value?
4. From the results of Problem 3 in the previous tutorial, use Givoni and Goldman equations to estimate HR after 5 minutes of work and then 5 minutes of recovery, and compare them to the actual values. How does the estimate of HR from Borg's RPE compare with the actual measured value?

5.8 LIFTING ANALYSIS

OBJECTIVES

- To collect data on static body joint angles during certain lifting postures
- To use biomechanics formulae to analyse the internal body forces when doing a lifting task
- To compare the forces and moments of a stoop lift versus a squat lift versus an erect stance
- To compare Chaffin et al. (2006) results with the NIOSH results and the data in Mital et al. (1997)

APPARATUS

Two briefcase boxes (see Web site) of 100 N and 200 N weight each	Video recorder and monitor
	Body joint markers with Sellotape
Video camera + playback facility	Balance scale
Anthropometer	Tape measure
Goniometer	Timber supports on floor for boxes
Wall boards marked with 100 mm squares	

TECHNICAL BACKGROUND

In regard to lifting, there has long been a debate about the squat lift versus the stoop lift (Burgess-Limerick et al. 1995) and some material on its biomechanics is covered by Chaffin et al. (2006). An attempt is made here to try to reproduce this in the laboratory and to use some biomechanics formulae to make calculations for comparison. Some of the body dimensions are rather difficult to obtain, but reasonable approximations are possible so that order of magnitude results are obtainable. It is also related to an actual manual materials handling task to help to add to the realism. For comparison use the NIOSH equation to get another idea of what the "safe" limits might be (Waters et al. 1993).

PROCEDURE

1. View a video of an actual lifting job in industry to show the complex angles and positions involved.
2. Select three participants to represent small, medium, and large members of the population. Measure stature and body mass.
3. Place markers at the centre of the ankle, knee, hip, elbow, wrist, and shoulder joints. Assume that the centre of mass of the upper body is 27% of the distance down from the shoulder joint marker to the hip joint marker. Presume that it acts vertically through this point, and mark this point on the participant. Estimate the position of L5/S1 (see figure in Chaffin et al. [2006]) at about 9 cm from the outer surface of the back and put a marker on it. Measure the distance from the centre of the hip joint to the centre of

the shoulder joint, and the distances from the shoulder joint to L5/S1, and from the hip joint to L5/S1 of each participant.

4. Position the load box on timber supports to provide clearance underneath for the fingers when picking it up, so that it can be held with the fingers underneath it.
5. Each participant then holds the box at each of these three postures:
   Stoop lift with the load just off the floor and close to the shins
   Squat lift with the load just off the floor and ahead of the knees with arms around the outside of the knees
   Erect (just hold the load at waist height, arms horizontal).
6. Video the participant while holding each posture for 10 seconds, and also measure the position of the centre of the load in each case from the centre of L5/S1 (h) and the centre of the upper body mass from L5/S1 (b) OR read off these values from observing the wall boards.
7. Repeat all the above with the other box after a suitable rest (at least 5 minutes).
8. Play back the tape in the camera to the monitor and use the pause button and the protractor on the screen to measure the body angles at: the knee joint (K), hip joint (Theta H), torso axis from vertical (T), upper arm from horizontal, lower arm from horizontal, and between the line from the hip to L5/S1 and the line from the shoulder to L5/S1.

REQUIREMENTS

1. A long laboratory report.
2. Calculate the approximate upper body mass of each participant. See table of Chaffin et al. (2006) for estimated mass of arms, head, and torso above L5/S1, which averages out at about 47.6% of the whole body mass.
3. Calculate the effective erector spinae force, and the compressive force and shear force at L5/S1 for each participant on each load with each posture, and calculate the moments and forces at the elbows and shoulders for all combinations—using the static, planar, and low back models of Chaffin et al. (2006).
4. Plot families of graphs for the forces against stature in one case, and against body mass in another one with the horizontal axis drawn to scale.
5. Examine the likely incidence of injuries for the three sizes of participant for the loads used, and compare these to NIOSH, Mital et al. (1997), and BS ISO 11228 (2007).
6. Outline what changes you would recommend to this lifting task, and why and how they would reduce the likelihood of injuries. Refer to data in the NIOSH guide or Chaffin et al. (2006).
7. Comment on the adequacy of the technique, procedures, and analyses used and their relevance to occupational activities. How do these results compare to those of Burgess-Limerick et al. (1995)?

REFERENCES

BS ISO 11228, 2007, *Ergonomics: Manual Handling*, British Standards Institution, London.

Burgess-Limerick, R., Abernethy, B., Neal, R.J. and Kippers, V., 1995, Self-selected manual lifting technique: functional consequences of the interjoint coordination, *Human Factors*, 37, 395–411.

Chaffin, D.B., Andersson, G.B.J. and Martin, M.B, 2006, *Occupational Biomechanics* 4th ed., Wiley, New Jersey.

Mital, A., Nicholson, A.S. and Ayoub, M.M., 1997, *A Guide to Manual Materials Handling* (2nd ed.), Taylor and Francis, London.

Waters, T.R., Putz-Anderson, V., Garg, A. and Fine, L.J., 1993, Revised NIOSH equation for the design and evaluation of manual lifting tasks, *Ergonomics*, 36, 749–776.

6 Systems Evaluations

Many of the exercises described in this chapter can be carried out by students working on their own, although working with others will help to achieve better results and a better learning experience. They all differ somewhat from the classical type of experiments such as those given in Chapter 5 and, in fact, they constitute a type of fieldwork in most cases. That is why it was deemed more appropriate to classify them as evaluations.

Initial material in this chapter relates to the Harmonising Education and Training Programmes for Ergonomics Professionals (HETPEP) requirement D.1 of systems theory. The first exercise helps to develop an understanding of the systems view among the students, and this point is accentuated by the next assignment, which is about how procedures contribute to the functioning of an organisation, as well as looking at individual work activities. These developments are extended by evaluating two workplaces from a systems point of view. However, experience has shown that students find it very difficult to view systems components from a functional viewpoint divorced from their physical role.

The other exercises are more concerned with health and safety issues, looking at aspects of system design where deficiencies can lead to errors and accidents and highlighting these as major contributors to accidents and injuries. In the case of the Error Cause Removal (ECR) task (see Section 6.5), the opportunity is taken to include also the issue of design for the impaired. The approach is intended to encourage the incidence of errors so that students will have something to think about, and an opportunity to devise improvements. In these ways the study deviates somewhat from Swain's original idea, and is a bit artificial, but that is also an inevitable part of the difficulty in trying to bring the "real world" into the academic environment.

Two of the exercises (Sections 6.5 and 6.6) are directly related to typical accidents, and a systems view is taken of how to minimise their likelihood of occurrence, their severity, and the gravity of the ensuing injuries to equipment or people. These can be related easily to countermeasures using ideas and techniques such as those propounded by people such as Willy Hammer, even though these do not lend themselves readily to laboratory work. In fact, these exercises should be supplemented by tutorial work suggested in such textbooks.

It will also be seen that this chapter relates to some of the issues of human reliability that have become such an important field in ergonomics work. However, these issues should be approached more by the use of probabilistic models and Design for Reliability techniques, which are not amenable to the forms of laboratory work described in this book. However, the two areas should integrate well with each other in the academic course as a whole.

PARTICULAR EQUIPMENT NEEDS

For the ECR investigation (see Section 6.5), the face validity will be better with a wheelchair for one person, a pair of crutches for the second, and a blindfold for the third. However, these are not strictly necessary as the difficulties can easily be imagined by most people.

6.1 SYSTEMS ANALYSIS

OBJECTIVES

- To deduce for a particular system what functions are required to turn its inputs into its outputs
- To determine which components carry out the functions
- To decide which functions can be eliminated or combined or simplified
- To decide which components should perform the functions

TECHNICAL BACKGROUND

Every system exists for a purpose (its objective), and uses inputs to achieve this by converting them into a set of outputs by means of various functions. These functions enable the conversion of inputs to outputs. Functions have their own inputs and outputs within the system, and they link to each other. Thus, the outputs of some functions are the inputs to others (Singleton 1974).

In designing a person–machine system, the functions are allocated to machines and/or persons at the function allocation stage. The way in which this is done critically affects system performance, and guides for assisting in these allocation decisions are provided by Shneiderman's Table (Appendix VIII) and Fitts' List. Existing systems can be analysed using these to determine whether or not functions have been allocated in the best possible way and, if not, how to improve it.

The results of a systems analysis can be presented conveniently in a block diagram such as Figure 6.1. Such a diagram depicts all the inputs, functions, components, and outputs, and shows how they are linked together. Critical examination of it can show where functions and/or components should be combined, substituted, simplified, or eliminated in order to improve the overall performance of the system. In other words, the functions can be allocated differently, or they can be performed by different components, or performed by new components not currently in the system. This process is normally enhanced by resorting to the Critical Questioning Technique (Appendix I), and by referring to Shneiderman's Table and Fitts' List.

CONVENTIONS

- Functions are described by one verb and one noun.
- Verbs are in the present tense and in active voice.
- The descriptions are abstract and *not* concrete to the situation (e.g., if person reads a form, function = "Extract info", or similar).
- All components involved in the function are listed in the block.

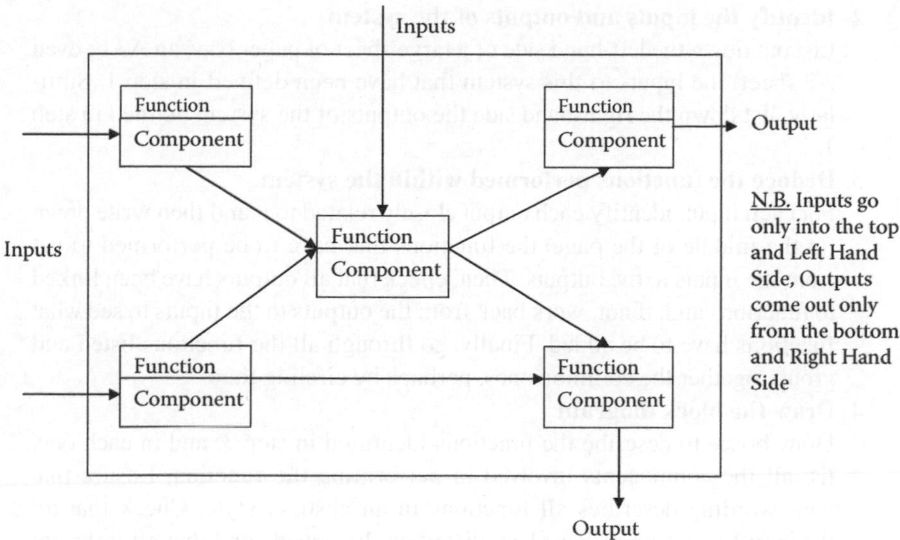

FIGURE 6.1 Example of a block diagram type of flowchart.

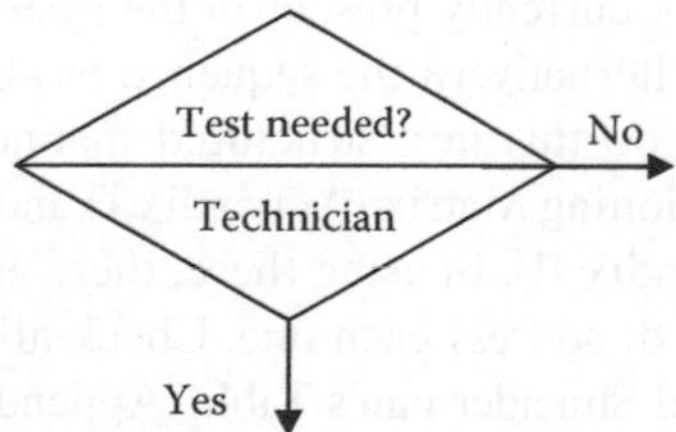

FIGURE 6.2 Example of a decision diamond.

- Decision "diamonds" (Figure 6.2) are used whenever a decision is required.
- Inputs and outputs are entities, *not* actions/events/destinations.
- The person making a decision is shown in the diamond.

PROCEDURE

Read thoroughly through the description of the hospital laboratory system given here. Then go through the steps in the mentioned order.

1. **Identify the system objective, and hence, its boundaries**
   What purpose does it serve and what service does it provide? This is the only way to decide where the system boundaries are, what entities are within the system, and what lie outside it. Remember that this has nothing to do with the physical walls of the laboratory or of the hospital.

2. **Identify the inputs and outputs of the system**
List out down the left-hand side of a large sheet of paper (say, an A3 or even A2 sheet) the inputs to this system that have been defined in step 1. Similarly, list down the right-hand side the outputs of the system defined in step 1.
3. **Deduce the functions performed within the system**
For each input, identify each output closely related to it and then write down (in the middle of the page) the functions that have to be performed to get from the inputs to the outputs. Then, check that all outputs have been linked to functions and, if not, work back from the outputs to the inputs to see what functions have to be added. Finally, go through all the functions listed and group together the common ones, perhaps by circling them.
4. **Draw the block diagram**
Draw boxes to describe the functions identified in step 3, and in each one, list all the components involved in performing the function. Ensure that your wording describes all functions in an abstract style. Check that all the inputs and outputs have been listed on the outside and that all links are shown.
5. **Critique this description of the system**
Examine whether or not functions or components or both can be combined, eliminated, simplified, substituted, changed, or reallocated. Perhaps new components not currently present in the system can be introduced to help this process. Alternatively, the sequence or method could be changed advantageously. To do this in a structured manner, use Konz and Johnson's Critical Questioning Matrix (Appendix I) and apply the Work Design Check-Sheet (Appendix II). In using these, there are often several answers to the question asked; address each one. Check allocations by referring to Fitts' (1951) List and Shneiderman's Table (Appendix VIII).
6. **Produce a new block diagram**
The desired end result is to come up with a better solution, so a diagram of how it will work is needed. Strictly speaking, this should then be subjected to further critical questioning and so on until no further improvement can be devised, but that would be too onerous to demand here.

REQUIREMENTS

1. A professional type of report.
2. A block diagram of the existing system, giving the information as set out in the style of Figure 6.1 but in more detail, on an A3 sheet. State the objective and ensure that there are at least 25 functions (not counting decision diamonds). Number them sequentially for ease of reference in points 3 and 4. Many functions utilise more than one component, so ensure that all components are listed in each function box.
3. Apply critical questioning in detail to the function of taking a specimen (e.g., blood sampling) from the patient for testing. List and number at least four variations of doing this, and deal with each one in each and every one

of the other boxes in the Critical Questioning Matrix (Appendix I). Make an A3 enlargement of the matrix and fill in the details in the boxes of one sheet only, including those in the left-hand column. Avoid vague generalisations when doing so.

4. Apply the Work Design Check-Sheet (Appendix II) to examine critically the functions performed in your block diagram. Which should be eliminated and why, which combined with which, and which simplified? List these out in a table on an A4 sheet divided into three: an upper part dealing with the eliminated functions, the next part for those combined, and the last part for those simplified. *Note:* Ensure that this is not confused with point 5.
5. Use Fitts' (1951) List and Shneiderman's Table (Appendix VIII) to examine critically the allocation of functions to components. Which components are wrong or unsuitable for the functions allocated to them (and why), what new components are needed (if any) and why, and which components should be eliminated or simplified or combined or replaced, and why. The emphasis must be on the appropriateness of the allocations. List these out in a table on an A4 sheet.
6. Summarise the aforementioned deliberations by drawing a new block diagram.
7. Include at least two A4 pages (or 600 words, where 5 letters = 1 word) discussing this work. Can any of the inputs or outputs be modified or eliminated to improve operations, are there likely to be knock-on effects of these proposals? What about the costs? What sort of problems can be anticipated in implementing them? What are the probable overall effects of them on this laboratory system, etc.? Do not rehash earlier material.
8. List numbered conclusions on the findings from this analysis.

REFERENCES

Fitts, P.M., Ed., 1951, *Human Engineering for an Effective Air Navigation and Traffic Control System*, National Research Council, Washington.

Sanders, M.S. and McCormick, E.J., 1976, *Workbook for Human Factors in Engineering and Design*, Kendall/ Hunt, Dubuque.

Sanders, M.S. and McCormick, E.J., 1992, *Human Factors in Engineering and Design* (7th ed.), McGraw-Hill, New York.

Shneiderman, B., 1987, *Designing the User Interface*, Addison-Wesley, New York.

Singleton, W.T., 1974, *Man-Machine Systems*, Penguin Books, Harmondsworth, Middlesex, U.K.

HOSPITAL LABORATORY

Specimens arrive in the laboratory via four principal routes. Doctors will send patients to the lab with orders for a test, or a series of tests, to be performed. If the test can be performed at that time, a laboratory assistant or technician will perform the test. If the test cannot be performed, an appointment is made for the patient to return, and an information pamphlet is given to the patient concerning the tests he or she will undergo.

In some cases, "inpatient orders" are sent to the lab requesting a laboratory assistant to come to a patient's room in the hospital and obtain a specimen (blood, urine, etc.). In other cases, the specimen is taken by a nurse on the patient's floor and just the specimen is sent to the lab with an order form requesting various tests.

After the specimen arrives in the lab, a laboratory assistant reads the orders to determine which tests are to be performed. He or she may have to call the requesting physicians if he or she cannot read their handwriting, or to notify them if there will be an inordinate delay in obtaining the results of the tests. The assistant then obtains the needed supplies, for example, slides, chemicals, test tubes, etc., and readies any required equipment. Some tests are very simple, whereas others are complex and require multiple steps. The procedures are specified in department laboratory manuals. In many cases, the tests are automated. A specimen is simply placed in the test equipment, and the results are read from various indicators on the machine or printed out by the machine. A time and materials card is filled out for each test and forwarded to Central Accounting for billing and inventory control. The results of each test are checked against a "norm chart" to determine if the test indicates any abnormalities. The results are transcribed onto laboratory report forms. One copy of the laboratory report is sent to the requesting physician, one copy goes to the Chart Room to be placed in the patient's hospital file, and one copy is placed in the laboratory's own files.

This more or less orderly process is often interrupted when a "stat" order is received. Stats are emergency orders that take priority over all other work and earlier orders.

Source: McGraw-Hill material from page 3 of Sanders, M.S. and McCormick, E.J., *Workbook for Human Factors in Engineering and Design*, 7th ed., ©1993, reprinted with permission.

6.2 PROCEDURAL ANALYSIS

OBJECTIVES

- To deduce for a particular system what procedures are presently carried out in it and what documents and records are created
- To determine which system components presently carry out the procedures, and create documents and records
- To decide which procedures, documents, and records can be eliminated, or combined, or simplified, or changed, or reallocated
- To assess what new procedures, documents, and records are required, if any
- To decide which system components should carry out the procedures and create the documents and records

TECHNICAL BACKGROUND

All organisations have systems and procedures to gather information, to record it, to transmit it between individuals and departments, and to make decisions based on it. These require documents and records to be compiled, and often, more than one person or department is involved in completing them. Similarly, copies are often distributed to other people or departments or both, and yet others extract information from these documents and the records made from them and may create new records or reports from them.

Over time, the needs of people or departments change so that some of the information (or an entire document or record) becomes redundant after a while. Similarly, a new item of information arises or a new need arises that fails to be gathered, recorded, and/or transmitted on a regular basis. Sometimes, information that should have been recorded is not, or the information goes to the wrong place or person, or is inaccurate owing to errors. In addition, it is sometimes the case that cumbersome procedures delay activities or introduce errors, and in other cases, decisions are made at too high or too low a level in the organisation or simply by the wrong person. Failures of these systems and procedures can cause major problems in the operations of organisations, and may result in accidents or injury if they happen within the safety system. However, they can also impair productivity, quality, and effectiveness. Often, such failures are ascribed to "a failure of communications". In addition to such issues, studies of the routes taken by information reveal the formal and informal organisation structure, and they can show shortcomings in that structure or the decision processes or both. Similarly, making changes in a procedure may have a major impact on the structure or on decision processes, and therefore, careful consideration of this possibility is required before anything is changed.

These activities can be examined in the same way as in some of the mechanistic task analysis techniques addressed in Chapter 3, with a series of special symbols joined in a way that reflects the information flows and connections between departments or functions. Undertaking such an investigation helps greatly to grasp better the systems approach, to develop a systems view of the organisation, and to highlight the need for good systems and procedures. We have not been able to determine who gave it the name used here in this context; it has been used in other contexts also. Similarly, we have not found the origin of the layout shown in the framework in Table 6.1, but the senior author was introduced to it at the University of Birmingham in the 1970s. It was described and developed in some detail by Mundel (1978), who labelled it Process-Chart-Combined Analysis, with a special set of symbols (some from American Society of Mechanical Engineers [ASME] 1972) and a checklist. However, it has not been well publicised despite its usefulness.

PROCEDURE

Examine the activities in a particular function of an organisation (in the university or outside), particularly where there are appreciable numbers of staff or a fair amount of processing of documentation, paper or electronic. There will be several activities on the campus that should be suitable. First, do a preliminary study to find out the general pattern of activities and the documents and procedures used. Then study the parts in detail.

REQUIREMENTS

1. A professional type of report.
2. Complete a Procedural Analysis Chart (laid out as in Table 6.1) describing the existing system: the operations and procedures performed and the docu-

TABLE 6.1
Framework of a procedural analysis chart for a hospital physician's activities

Forms/ documents	Subdivisions or departments								Operations performed
	A & E	Ward staff	Laboratory	X-ray	Surgery	Administration	Clinic	Specialists	
A: Doctor order									1. Examine patient
B: Patient record									2. Record symptoms
C: Lab form									3. Check records
D: X-ray request									4. Decide on test
E: Lab report									5. Write out order
F: X-ray plate(s)									6. Consult specialist
G: Clinic request									7. Contact family
…………….									………………
X: Surgery record									19. Study test data
Y: Special request									20. Write up case

Note: Symbols from the next page are inserted in the appropriate columns, with the requisite letter or number codes, and joined by flow lines to the related symbols in other columns to show how activities and record systems link together, or do not, as the case may be. It will also show how and where information is disseminated across the organisational subdivisions or departments, where it is available, who uses it, what is missing, etc.

TABLE 6.2
Procedural analysis symbols

SYMBOL	Meaning
❐X B	Document or file created in X copies (where X is as appropriate). B refers to the code in the left-hand column of the Procedural Analysis (P.A.) Chart (see Table 6.1) for a form or document
③	Operation performed on the record (e.g., computation made, or extra information added). 3 refers to a description of the operation in the right-hand column of the P.A. Chart (Table 6.1)
↑	Arrow means "return to file" or "save" or similar action
□	Inspection: means the correctness of information is checked by comparison with other source(s) of information
- - - - -	Broken line shows a connection to another source on the P. A. Chart
↵	Information takeoff: shows the information extracted and entered elsewhere, or used separately by someone. The point indicates another symbol or a parallel chart to which it is going. Use the broken line as shown earlier.
⊗	Disposal: indicates the record or copy is deleted or destroyed
⇨	Movement: indicates that the record changes location (e.g., to another directory) without any change in it
D	Delay or Temporary File: indicates records that are waiting to be worked on (e.g., in an in-tray or in an e-mail folder or on a storage medium)
▽	Permanent Storage: indicates that the record is put into a folder, file, or directory, organised in a formal fashion
Ø	Item Change: shows change occurs in the item recorded
⌗	Gap: shows activities not pertinent to the study, not itemised in detail
▤	Existing Record or Document: one that was created in a previous section

Note: Use others as needed for specific applications.

All symbols used are taken from the symbols and wingdings in Microsoft Word. Five are the standard ASME Symbols used for process charts, and the others all differ from those given in the book.

Source: Nofsinger, *Motion and Time Study*, 5th edition, ©1978, p. 210. Adapted with permission of Pearson Education, Inc., Upper Saddle River, NJ.

ments and records that are used and created, using the Procedural Analysis Symbols given in Table 6.2.

3. Apply the Procedural Analysis Checklist to the chart to examine it critically, and also use the Work Design Check-Sheet (Appendix II). List the results of your critique of the chart.
4. Devise improvements to the systems and procedures, and revisions to the documents and records, by eliminating, combining, simplifying, or real-

locating systems and procedures or documents or records, or by adding new systems and procedures or documents or records, or using hardware or software to help.

5. Draw up a new Procedural Analysis Chart to depict your improved system design, and demonstrate clearly the effects of your improvements.

REFERENCES

ASME, 1972, ASME Standard: Operation and Flow Process Charts, American Society of Mechanical Engineers, New York. www.asme.org

Mundel, M.E., 1978, *Motion and Time Study* (5th ed.), Prentice-Hall, Englewood Cliffs, NJ.

PROCEDURAL ANALYSIS CHECKLIST

1. Is each step really necessary? If not, eliminate it.
2. Is each step at the ideal place in the sequence? Where should it be?
3. Does each step have a reason for being by itself? Can it be combined with others?
4. Is each step as easy as possible? How can it be made so?
5. Does each record have a real purpose? What is it? Is the record necessary? Can it be eliminated, combined with another record, or replaced by a copy of another record?
6. Does each file or folder have a unique purpose? Is there duplication? Are there excessive files? Is it filed by the subject used to enter files? What is the manner of its use?
7. Is the record finally destroyed? Should it ever have been originated? What is its purpose?
8. Does information go from one record to another? Are more copies needed? Are all information take-offs/readings necessary? Which have priority? Which should have priority?
9. Are all copies being used equally? Can the load be shared to speed up the procedure?
10. Does someone sign all copies? Can this be avoided? How? Signers are often busy people.
11. Is there excessive checking of actions or records? Can it be reduced? How?
12. Where is the best place to check? What is the risk of a mistake, and its effects?
13. What would happen if the record were lost? Are backups needed?
14. What equipment or software might help the job?
15. Does one person handle too much of the procedure? What happens if he or she is absent?
16. Are as many steps as possible given to the lowest classification of personnel?
17. Can the travel or transmission of records be reduced advantageously?
18. Can the record be kept in action, out of in-trays or inboxes or in "pending" locations?
19. Does the information arrive in a timely fashion so that it may be acted upon adequately?
20. Has the information been reduced to an understandable form?

21. Is the information clear, accurate, and reliable?

Source: Nofsinger, *Motion and Time Study*, 5th edition, ©1978, pp. 212, 214. Adapted with permission from Pearson Education, Inc., Upper Saddle River, NJ.

6.3 OFFICE WORKPLACE EVALUATION

OBJECTIVES

- To apply the lecture material to the process of evaluating a workplace chosen by the student, such as a person using familiar software on a PC.
- To take a systems view, that is, evaluate the total combination of person, chair, desk, keyboard, screen, workspace, and software.

TECHNICAL BACKGROUND

There is often a tendency to examine office workplace problems separately, such as a chair, desk, monitor, keyboard-positioning problem, and so on. The principle to be applied here is that these are all parts of a system that has to be looked at as a whole. The aim is to design some aspects of a study and to carry out an investigation of dimensional aspects. Therefore, all the evaluations should be carried out on the one workplace. The study is assumed to be done by a single person and, to keep the requirements within practical bounds, examinations of noise, lighting, people traffic, social environment, etc., have been deliberately omitted. However for a class field study, they would form very important parts (see Section 6.4).

PROCEDURE

Consult Wilson and Corlett (2005) for detailed advice on the appropriate methods, procedures, etc. There are three separate parts to the procedure as follows:

1. **Subjective Measures:** Establish which to use (and explain in some detail why), drawing up measurement scales for various aspects of each of five of the relevant attributes (but not chair features, dimensions, or questionnaire). Each must be for a different type of attribute, that is, only one on distances (say), only one on comfort, and so on. Put an example of each one separately in the report, and justify and explain in some detail the reasons for the choice of each scale and its features but not of the attribute. Say what will be used; not "maybe", or "could be", or "might be", or "we should consider". How do they meet the requirements specified in lectures and references?
2. **Design an Experimental Study:** Describe specifically what exactly to use for carrying out an experiment on this particular workplace (but do not carry it out). Give the dependent variables and independent variables, the measures to use for each of these, and the design to use for the experiment (e.g., number of levels for each and what they are physically and why, orders (why), balancing measures (why), what are held constant and why, type of design and why, covariates (if any) and why, etc.). Give reasons and

justifications for the choices. The points made must relate to this specific workplace with actual variables, actual levels, actual measures, and so on, and not "maybe's" or "could-be's" or "might-be's". What analysis will be used on the data and why? From what is submitted, an undergraduate must be able to go out and perform the experiment.

3. **Suitability of Workspace Dimensions:** Do not repeat material from parts 1 or 2 just mentioned.
   a. Examine dimensions that are relevant; measure them, present them in a table, and contrast them with other data (Kroemer, 1993 including those in the article) just by comparing numbers.
   b. Measure five people from the population available to use the workplace for body dimensions 2, 3, 5, 6, 7, and 11 from the anthropometry part of Sanders and McCormick (1992). Present them in a table with means and standard deviations. Summarise both sets in one table.
   c. Assess the match of this population to these data sets by doing *t*-tests on each of the dimensions. For this reason, try to use a job involving males but, if this is not possible, work from the U.S. female data. Use coefficients of variation given by Pheasant and Haslegrave (2006) to estimate standard deviations where there are none.
   d. Assess how the study participants match the dimensions of the workplace measured in step a. Do this by simple comparisons of dimensions.
   e. How would the workplace described in step a cope with participants at either end of the size range? Use coefficients of variation to estimate standard deviations (where none are available), and then use the *t*-distribution to get estimates of the sizes at the extremes (1st and 99th percentiles). Then compare values and look at proportions outside the limits available at the site.
   f. Identify the deficiencies in all the workplace dimensions (and features, if appropriate).
   g. Give proposals for improving the workplace relative to step f, and explain how and why they will help. (See ILO 1996 and Yu et al. 1988 for general guidance on problems and approaches [but not for reproduction of parts of it as your answer!] and the Web tool mentioned in the reference list.)

REQUIREMENTS

1. A professional type of report with suitable tables for each part

REFERENCES:

ILO, 1996, *Ergonomic Checkpoints*, International Labour Organisation Publications, Geneva.

Kroemer, K.H.E., 1993, Fitting the workplace to the human and not vice versa, *Industrial Engineering*, March, 56–61.

Pheasant, S. and Haslegrave, C.M., 2006, *Bodyspace: Anthropometry, Ergonomics, and the Design of Work* (3rd ed.), Taylor and Francis, London.

Sanders, M.S. and McCormick, E.J., 1992, *Human Factors in Engineering and Design* (7th ed.), McGraw-Hill, New York.

Wilson, J.R. and Corlett, E.N., 2005, *Evaluation of Human Work* (3rd ed.), CRC Press, Boca Raton, Florida.

Yu, C.-Y., Keyserling, W. and Chaffin, D. B., 1988, Development of a work seat for industrial sewing operations: results of a laboratory study, *Ergonomics*, 31, 1765–1786.

http://www.ergonomics.ie/mirth.html provides access to aids developed under the EU funded MIRTH (Musculoskeletal Injury Reduction Tool for Health and safety) project.

6.4 FACTORY WORKPLACE EVALUATION

OBJECTIVES

- To apply the lecture material to the evaluation of an industrial workplace by a team
- To take a systems view, that is, to evaluate the total combination of person, stool, bench, environment, tools, parts, lighting, noise, postures, and materials handling

APPARATUS

Video camera	Still camera
Noise meter	Light meter
Tape measure	RULA sheets
Thermometers (wet and dry bulb, globe)	Anthropometer
Air velocity meters	Smoke generator

TECHNICAL BACKGROUND

An evaluation, such as would be carried out by consultants, must examine all the interlocking features of the person's work situation; see the work by Harris (1986) for guidance. To be able to accomplish this within a limited time, it is necessary to do this by means of a team approach, which will also give practice in working as part of a team. One of the aims of HETPEP is to develop an "ergonomics viewpoint", which is essentially a systems view, and studies of this type are an important part of forming that view as opposed to a physiological view, or a psychological view, or an occupational hygiene view.

The purpose is to integrate material covered in the course into a systems view of what is happening in a particular factory floor work situation, and to decide if the combined effect of the various factors makes for an acceptable or an unacceptable work situation (ILO 1996; Wilson and Corlett 2005). It also provides an opportunity to get a feel for the way that the various factors interact with one another, and to become familiar with working with factory floor people.

PROCEDURE

Form teams of three.

Visit a manufacturing plant (preferably) and examine as many of the work aspects as can be managed in one morning or afternoon (or a day, if that is possible). Each team member takes on specific roles and, later, an overall assessment is compiled by the team. Examine postural problems with RULA, measure the thermal environment (wet bulb, dry bulb, globe thermometer if there is a significant amount of radiant heat), ventilation (air velocity, flow patterns), lighting (levels and glare), noise (preferably with an individual dose meter), anthropometrical issues, materials handling, physiological demands, biomechanics, workplace layout, and space constraints—whatever is possible or relevant. See the work by Pheasant and Haslegrave (2006) and the Web tool mentioned in the reference list for suggestions.

If possible, make a short video of the workplace tasks and any nontrivial manual materials-handling activity. Draw up a two-handed flow process chart of what the person does in one work cycle and extract some broad time data, at least. Take a number of still photographs to show the salient features of what the workers do.

Take measurements on the relevant body dimensions of the worker, bench, stool, workplace layout, parts, tools, masses, etc.

REQUIREMENTS

1. A professional type of report.
2. Ensure that you have assessed as many ergonomics aspects of the workplace as are relevant.

REFERENCES:

Harris, P., 1988, *Designing and Reporting Experiments*, Open University, Milton Keynes, U.K.

International Labour Office, 1996, *Ergonomic Checkpoints*, ILO Publications, Geneva.

Pheasant, S. and Haslegrave, C.M., 2006, *Bodyspace: Anthropometry, Ergonomics, and the Design of Work* (3rd ed.), Taylor and Francis, London.

Wilson, J.R. and Corlett, E.N., 2005, *Evaluation of Human Work* (3rd ed.), CRC Press, Boca Raton, FL.

http://www.ergonomics.ie/mirth.html provides access to aids developed under the EU funded MIRTH (Musculoskeletal Injury Reduction Tool for Health and safety) project.

6.5 ERROR CAUSE REMOVAL FOR DISABLED

OBJECTIVES

- To apply Swain's procedure to improve a real-world problem, in a nonmanufacturing situation
- To do this by analysing errors from the need for mobility of disabled students on campus
- To see how well the technique works, especially in a nonmanufacturing situation

TECHNICAL BACKGROUND

Today, due to building regulations, there is an emphasis on enabling people with disabilities to participate in normal activities as much as possible. Universities are trying to address these problems on campus by means of access committees. In this context, a student can be seen as the equivalent of a factory worker and so can perform as a member of an ECR team, stretching the point of Swain (1973) somewhat. However, the important point is to gain experience in identifying errors that are induced by systems and building design so that such errors can be eliminated or prevented by better design, especially when the limitations of users are addressed. It can also demonstrate that relatively inexperienced observers can make significant improvements to reduce opportunities for error.

The task to address is the risk of error caused by the mobility difficulties of people with disabilities such as blindness, wheelchair use, and confinement to crutches. It is very difficult to provide sufficient error examples to study, in a low risk environment, readily available on campus. The requirement is to examine the error situations that arise with such people when they try to perform very ordinary tasks as discussed in the following section.

AN EXAMPLE OF AN ACTIVITY TO INVESTIGATE

The activity starts off with any one of the students sitting at a table in the cafeteria. From there, the student goes over to the serving counter, goes up to the servery, buys a chocolate bar, fishes out coins and pays for it, emerges into the sitting area, crosses it towards the exit on the door side, and maybe down some steps to the door. Then the route goes across the campus to the building where students normally spend their time (say), into it and up some stairs (if possible) and along some passages. The blind student then goes through the door into a lecture room. The person on crutches goes by a different route and takes the lift (elevator) to an upper floor and along some passages into the same lecture room. The wheelchair person takes a route on a system of paths to the front entrance, down the corridor to the lift; then the route is the same as before. Each student takes a seat in the front row on the far side of the room. All this takes place between quarter to the hour and the hour on a cold, rainy, and windy day when there are many people moving about between rooms.

PROCEDURE

Divide into ECR teams of three; each concentrates on (and writes up) a different type of disability. To simulate blindness, one student should be blindfolded. A wheelchair and crutches should be borrowed in order to experience the other situations for real, if possible. Identify System Errors (SEs) that occur at present, Error Likely Situations (ELSs; not necessarily seen, but obviously potential errors), and Accident-Prone Situations (APSs; use your imagination). These are defined as follows (Swain and Guttman 1983):

SEs: Out-of-tolerance actions in which the human exceeds (or is likely to exceed) some limit of acceptable performance; the limits are defined by the parameters of the system itself.

ELSs: The system makes demands that are not (or are not going to be) compatible with human capabilities and limitations in regard to their perception, information processing, decision-making abilities, and so on; that is, these errors arise because of limits set by humans rather than the system.

APSs: Situations that foster human errors that are likely to result in injury to people, or damage to equipment or facilities.

Walk the route at the time specified so as to address the real situation. Note along the way the various classes of problems involved and design features giving rise to them. Devise possible changes to systems, products/equipment, or procedures that you consider likely to reduce the probability of such system failures that occur when people with disabilities carry out this set of activities. Consult the work by Norman (1988) for further ideas, and refer to the appropriate building regulations.

REQUIREMENTS

1. A professional type of report.
2. Use a sketch to show the route taken, and mark the features and hazards on it with a number. Remember the effects of other students moving about, many coming the other way or across the intended path or around corners. Also, be on the alert for other forms of traffic (vehicles, bicycles, trolleys, etc.).
3. Draw a table on A3 sheets (in landscape layout) with columns to identify the number of the point on the sketch, the type of problem (SE, ELS, or APS), description of system features that produce the problem, how they give rise to the problem, recommended changes in design or procedures, impediments to these changes, and measures to overcome these impediments. Cross-reference these points to the text and summarise the problems and the features causing them. Remember that simple errors such as losing one's way or going an unnecessarily long way or other similar simple difficulties should be eliminated. Reasons for impediments may be costs, vested interests, habits, or inertia.
4. Discuss thoroughly the suitability/difficulties of the technique, the quality of the proposals, the alternatives to these proposals that might be suitable (especially if the study were done again), and so on. This should be at least two A4 pages long.

Note: The emphasis should be on errors and accidents, *not* on access as such.

REFERENCES

Building Regulations

Norman, D.A., 1988, *The Psychology of Everyday Things*, Basic Books, New York.

Swain, A.D., 1973, An error cause removal program for industry, *Human Factors*, 15, 207–221.

Swain, A.D. and Guttman, H.E., 1983, *Handbook of Human Reliability with Particular Emphasis on Nuclear Power Plant Applications*, National Technical Information Service, Springfield, Virginia.

6.6 HAZARDS OF SLIPS/TRIPS/FALLS

OBJECTIVES

- To examine problems of slips, trips, and falls
- To gain experience in a technique for studying these mishaps
- To look at measures to reduce them or the severity of the results from them

TECHNICAL BACKGROUND

Health and Safety authorities produce figures that show that slips, trips, and falls constitute a significant proportion of occupational accidents (see Davis 1985; Ridd and Manning 1995). Haddon's (1973) ten countermeasure strategies (Table 6.3) provide a means of reducing such accidents, or at least their adverse effects. His concern was those phenomena in which "energy is transferred in such ways and amounts, and at such rapid rates, that inanimate and animate structures are damaged". The underlying viewpoint is that too much energy, or an ordinary amount of it in the wrong place, lies at the core of many forms of accident. The aim is to examine the system in order to analyse options, strategies, and their cost, so as to maximise the payoffs. His approach is to reduce energy levels, to provide some forms of containment or absorption of the energy, or to reduce the adverse effects of the release of this energy. Ideally, such an evaluation should be undertaken at the design stage.

The original application was for situations involving harmful interactions with people and property of things such as projectiles, moving vehicles, and ionizing radiation. However, his approach is applicable to slips, trips, and falls, that is, people on the move. The technique provides an instructive introduction to a common form of accident and injury, caused by unnecessary demands on people or inadequate design of their environment.

PROCEDURE

Conduct the study in and around the building in which the students normally study (or another suitable one nearby with which they are familiar), its corridors and stairs, and all its various entrances and approaches. The area studied should include pathways leading to the building, the paths from the car park, and any relevant stairways both inside and outside. It is important to take note of opposing streams of human traffic, particularly at peak times, and other traffic such as vehicles, bicycles, trolleys, garbage removal, gardening activities, etc.

Divide the whole area into several subareas and identify all the ways that slips, trips, and falls can occur in each area. For each, specify the types of injuries that can result, list Haddon's principles (often more than one) that apply to each injury (using numbers only), and then state the measures or hardware or both (or changes to existing hardware) to be used to implement Haddon's strategies to eliminate or reduce

TABLE 6.3
Haddon's countermeasure strategies

Number	Strategy
First	Prevent initial marshalling of the energy: Do not generate thermal, kinetic, or electrical energy; avoid concentration of chemicals; avoid pressurised containers; store objects at floor level; neutralise toxicants and prevent their production.
Second	Reduce the amount of energy marshalled: Reduce concentrations of chemicals; reduce storage heights; reduce vehicle speeds; limit the quantity and manufacture of toxicants; reduce the pressure in containers; prohibit running in corridors.
Third	Prevent the release of energy: Prevent the discharge of electricity; fit railings to the sides of bridges; fit safety catches to hunting weapons; fit interlocks; spray sand onto icy surfaces to prevent slipping; fit safety harness to workers at a height.
Fourth	Modify the rate of spatial distribution of release of energy: Reduce the slope of ski runs for beginners; use slow burning rate explosives; fit arresting gear on aircraft carriers; use fire-retardant clothing; supply parachutes.
Fifth	Separate the susceptible structure in space or time from the energy being released: Separate pedestrians from vehicular traffic; place electric power lines out of reach; employ traffic overpasses and underpasses.
Sixth	Separate the susceptible structure from the energy being released by interposing a material barrier: Use safety goggles, helmets, gloves, electrical, or thermal insulation; fit fire doors; screen patients from x-rays.
Seventh	Modify the contact surface or subsurface of the basic structure that can be impacted: Eliminate or round off or soften corners, edges, and points where people can come in contact; make toys less harmful; provide automobile interior crash padding; provide "breakaway" roadside poles that yield gently on impact.
Eighth	Strengthen the living or nonliving structure that might be damaged by the energy transfer: Set tough codes for resistance to fires, floods, or hurricanes; provide vaccines against diseases; provide vehicle impact resistance and crush zones; require "preseason conditioning" of athletes.
Ninth	Move rapidly to detect and evaluate damage and to counter its continuation and extension: Generate emergency signal, transmit it, evaluate it, and follow up; use alarms, flood warnings, sprinkler systems, SOS calls; stop haemorrhaging; despatch ambulance and/or fire crew; give mouth-to-mouth resuscitation.
Tenth	Stabilise the process after intermediate and long-term reparative and rehabilitative measures: Return the system to its pre-event status and/or undertake rehabilitative measures; provide plastic surgery; develop less expensive automobile repair methods; fill in Chernobyl reactors with concrete.

See also Appendix B for four case studies to get a more detailed understanding.

Note: Although the strategies are in rank order of increasing loss reduction, there is no logical reason to start always at the lowest or lower rankings.

Source: Adapted from Haddon, W., 1973, Energy damage and the ten countermeasure strategies, *Human Factors*, 15, 355–366.

these energy-related accidents and their effects—for each principle in turn. Consult the work of Reese (2001) for suggestions.

REQUIREMENTS

1. A professional type of report.
2. Use a sketch to show the places where the slips, etc., are likely to occur, and mark the features and hazards on it using numbers. Remember to consider the effects of weather, activities of cleaning and janitorial staff, human traffic along paths and corridors and around corners, and other forms of traffic.
3. Draw up a table on A3 sheets (in landscape layout) with columns to identify the number of the point on the sketch, the type of problem (slip, trip, or fall), description of system features that produce the problem, how they give rise to the problem, recommended changes in design or procedures, impediments to these changes (and measures to overcome these impediments), methods to protect people and equipment, ways to reduce the effects of these mishaps, and so on. Cross-reference these points to the text in the narrative. In the text describing the findings, summarise the problems and the features causing them.
4. Examine thoroughly the suitability or difficulties of Haddon's strategies, the quality of the proposals, and alternatives to these proposals, especially if the evaluation were done again. This should amount to at least two A4 pages.

REFERENCES

Davis, P.R. (Ed.), 1985, Special issue on Slipping, Tripping and Falling Accidents, *Ergonomics*, 28, 945–1085.

Haddon, W., 1973, Energy damage and the ten countermeasure strategies, *Human Factors*, 15, 355–366.

Reese, C.D., 2001, *Accident/Incident Prevention Techniques*, Taylor & Francis, London.

Ridd, J. and Manning, D.P. (Eds.), 1995, Special issue on Slipping, Tripping and Falling Accidents, *Ergonomics*, 38, 193–259.

6.7 HAZARD ANALYSIS ON TRANSPORTATION AND HANDLING

OBJECTIVES

- To examine some hazards of transportation and handling in a work area
- To gain experience in Brown's technique for studying hazards
- To look at measures to reduce or eliminate them and the severity of the results

TECHNICAL BACKGROUND

Brown (1976) developed a procedure for collecting data on potential hazards and for analysing their causes and possible countermeasures. The first part consists of performing a General Hazard Analysis (GHA) on each hazardous activity in the organisation or area under study. A brainstorming session is initiated to think up a wide range of possible hazard scenarios and their causes. Each hazard is then rated in purely qualitative terms on each of the following four aspects (terms modified slightly to facilitate coding and clarity):

Hazard severity:	nuisance (N), marginal (M), critical (C), catastrophic (Cat)
Probability of occurrence:	unlikely (U), probable (P), considerable (C), imminent (I)
Cost if it happens:	prohibitive (P), extreme (E), significant (S), nominal (N)
Action to take:	defer (D), analyse further (A), immediate response (I)

The results from the GHAs are summarised into a matrix. This is then examined in detail to whittle the hazards down to those that are technically and financially feasible in a reasonable length of time and to identify the most urgent ones. This step is followed by a Detailed Hazard Analysis (DHA) on specific problem areas identified by the GHA. In order to perform the DHA, task analysis is performed on the activity in question, and then the ways in which hazards can occur from these tasks are spelled out. The results are recorded on the DHA form in a matrix arrangement. Wherever a task element and a hazard element occur together, the matrix cell is marked with an X.

All of the X-cell situations are then examined, again with brainstorming, to devise varieties of countermeasures that may require equipment or organisational innovations. Then a new DHA is drawn up, as a first revision to the initial design. In this second DHA, some of the cells marked X will have had the hazard eliminated, in which case the X is replaced with an E. In other cases, it may only be reduced so it is marked with an R; in yet other cases, the hazard level may have increased, so it is then marked I. The R and I can be used with a subscript to indicate different levels of reduction or increase. Also, of course, in some cases the hazard may remain unaffected, so the X will remain. If any hazards remain after the second DHA, the process is repeated until some irreducible minimum is reached.

Good knowledge of ergonomics is crucial for such a study; the experimenter should have sufficient awareness of potential areas of human error, or overload, or system failure. Thus, it should not be undertaken until the students have progressed some way into the course. It lends itself very much to field study work, but such a proposal may not be sympathetically received by employers, and adequate on-campus examples can usually be found.

PROCEDURE

1. Locate appropriate hazardous activities, preferably in areas where several people are present, such as students working in the engineering workshop, doing materials-handling work in the university goods-receiving area, or in a chemistry laboratory, physics laboratory, or in the campus kitchen.
2. Preferably assign separate student groups to study different activities. Whatever is studied, it is essential that the concerned students visit the site to examine the situation thoroughly while the activity is being performed by the usual people wherever possible.
3. Perform a GHA (Table 6.4).

TABLE 6.4
General Hazard Analysis (enter a letter code in each rating column)

Area assessed:	**DHA No.**
	Sheet of
	Student:
Activity performed:	**Group:**
	Present-proposed system (circle one)

Description of hazards that arise in activities in the area studied	Hazard severity	Probability it occurs	Cost if it occurs	Action to take

Source: Brown, David B., *Systems Analysis and Design for Safety*, 1st ed., ©1976, p. 47. Adapted by permission of Pearson Education, Inc., Upper Saddle River, NJ.

4. Choose the most hazardous-looking job in the area and perform a DHA (Table 6.5) on it.

One possible example is as follows: A student goes to the Engineering Workshop steel store where he or she has to remove a 4 m long bar of 25 mm diameter (mass about 16 kg) from the rack where bars of different sizes are stored horizontally in a jumbled manner. He or she has to get it into the mechanical hacksaw to cut off a length of 1 m; it is then taken out into the passage, goes along the passage and into the workshop proper, where it is taken to a lathe and loaded into the chuck. There it is machined and, when this is finished, it is removed and put onto a carrier.

REQUIREMENTS

1. A professional type of report.
2. Draw up the GHA (Table 6.4) for the set of activities.
3. Draw up DHA (Table 6.5) for the jobs that arise from your analysis.
4. Devise appropriate countermeasures (Hammer and Price 2000), and produce a GHA and DHAs for the proposed improved system.
5. In your appendices, summarise the hazards and the features causing them.
6. Examine thoroughly in one of the appendices the suitability or difficulties of Brown's procedure, the quality of the proposals, and possible alternatives to these proposals, especially if the investigation were done again and/or some money was spent on it.

REFERENCES

Brown, D., 1976, *Systems Analysis and Design for Safety*, Prentice-Hall.

Hammer, W. and Price, D., 2000, *Occupational Safety Management and Engineering*, Pearson, Upper Saddle River, NJ.

6.8 SHIFT WORK SCHEDULING

OBJECTIVES

- To gain experience in designing a shift work system
- To familiarise students with Knauth's approach to the design of such systems
- To acquaint students with various ways of ameliorating shift work ill effects

TECHNICAL BACKGROUND

The disruption of circadian rhythms due to night work is well known, and a variety of work schedules have been tried in order to ameliorate the ill effects. Knauth (1993) provides a clear plan for designs that will meet many concerns of ergonomists, and Monk and Folkard (1992) give guidance on ways to help shift workers.

Traditionally, these studies have been concerned with industrial workplaces, whereas this application is confined to a clerical work environment. This different setting poses slightly different questions and will be a novelty for many people even

TABLE 6.5
Detailed hazard analysis **For GHA no.____**

Area assessed:	DHA No.
	Sheet of
	Student:
Activity performed:	Group:
	Present-proposed system (circle one)

Task element nos.	Hazard element numbers and their state (X,E,R, or I) (draw in columns as needed)	Hazard element nos. and descriptions

Source: Adapted from Brown, D., 1976, *Systems Analysis and Design for Safety*, Prentice-Hall.

though in the United States and Canada it is quite normal for university libraries to be open on a 24-hour basis. It also raises some social issues that are often ignored in industrial situations.

With the advent of the system of working three or four 12-hour shifts per week, an alternative scheme can be drawn up and compared for advantages and disadvantages (see Kogi and Thurman 1993, special issue of International Journal of Industrial Ergonomics [IJIE] 1998). Similarly, questioning the staff concerned can add a useful extra dimension to the learning experience.

PROCEDURE

Prepare a proposal to address the following situation:

A 24-hour three-shift system is to be introduced, 7 days per week, 365 days per annum (including public holidays), for the operation of the university library. Establish suitable criteria for typical library staff with regard to their work hours, number of consecutive nights of work, amount of free time and free days, socialising needs, sleep needs, and the work physiology criteria, and then compile a workable plan. Examine also the specific health measures to be incorporated into the running of the plan. Remember that the normal attendance hours for such people are about 35 per week, so this or some close variation of it is required, at least on average. The normal day is assumed to be 09.00 to 17.00, but library staff tend to have nonstandard work schedules.

Include summer holidays in some way because the needs are less in June through September and for the weeks around Christmas or New Year. Note that leave is supposed to amount to 28 working days per annum over and above public holidays, and assume that these must include one continuous break period of 23 days. Ensure that total vacation time is not less than those for staff not on shifts. How successful do you expect your plan to be in minimising the problems of shiftwork?

Assume that there are currently 24 full-time staff who deal directly with the public one way or the other for the present level of service. Design it using the information provided by Knauth (1993), Kogi and Thurman (1993), and Monk and Folkard (1992) and give reasons for choices.

REQUIREMENTS

1. A professional type of report.
2. Charts to show the main features of the plan; specify and justify the design criteria used.
3. How do the hours work out each week, what are the staffing levels over the 24 h and/or week, what is the level of service to students, etc.?
4. Show the numbers, for example, by a small table with allocation of teams to days of the week over the full cycle and allocation of teams for each shift time in each day of each week. Use your imagination in designing the plan.
5. Ensure that the report includes the following: design criteria, details of operation, critique of the proposal, tables of shift regime, and hours worked per annum and per week. Say what to do and why, not could do or might do; that is, take a definite position rather than talking in generalities such as committees often do.

6. Compare alternative shift systems for this application.

REFERENCES

IJIE, 1998, International Journal of Industrial Ergonomics special issue on Shiftwork.

Knauth, P., 1993, The design of shift systems, *Ergonomics*, 36, 15–28.

Kogi, K. and Thurman, J.E., 1993, Trends in approaches to night and shiftwork and new international standards, *Ergonomics*, 36, 3–13.

Monk, T.H. and Folkard, S., 1992, *Making Shift Work Tolerable*, Taylor and Francis, London.

Some Equipment Suppliers*

ANTHROPOMETRIC INSTRUMENTS

Holtain Ltd, Crosswell, Crymych, Dyfed SA41 3UF, U.K.
Harpenden body callipers, skinfold callipers, etc.

Siber Hegner Maschinen AG, Wiesenstrasse 8, CH-8022 Zurich

BICYCLE ERGOMETERS

Cardiac Science Corporation, 3303 Monte Villa Parkway, Bothell, WA 98021, USA www.cardiacscience.com
Quinton bicycles, electrocardiology, etc.

Collins Medical, 220 Wood Road, Braintree, MA 02184-2408, USA www.collinsmedical.com
Pedal-mode ergometer

MONARK Exercise AB, S-432 82 Varberg, Sweden
Monark models

DATA LOGGERS

Grant Instruments (Cambridge) Ltd, Shepreth, Cambridgeshire SG8 6GB, U.K. www.grant.co.uk
Grant 1000 Series Squirrel, connects to pickups, e.g., for earlobe to detect HR
Rikadenki Kogyo Co Ltd, No. 17-11, Kakinokizaka, 1-Chome, Meguro-ku, Tokyo 152

ENVIRONMENT RECORDERS, E.G., FOR FIELD STUDIES

Onset Computer Corporation, 470 MacArthur Blvd, Bourne, MA 02532, USA
www.onsetcomp.com/products/3654 temp.html
HOBO 4-Channel System, e.g., for noise, temperatures, other

FORCE AND TORQUE MEASUREMENT

Mecmesin, Newton House, Spring Copse Business Park, Slinfold, West Sussex RH13 7SZ, U.K.
www.mecmesin.com
Mecmesin Force Meters

* Up-to-date information is available at CRCpress.com/e_products/downloads.

GENERAL PURPOSE EQUIPMENT

Lafayette Instrument Co, ByPass 52 & N. 9th Street, Lafayette, Indiana 47902, USA
Large variety of timers, logic units, testing devices, etc.

GRIP STRENGTH TESTERS

Lafayette Instrument Co, ByPass 52 & N. 9th Street, Lafayette, Indiana 47902, USA
Manual type

MIE Medical Research Ltd, 6 Wortley Moor Road, Leeds LS12 4JF, U.K.
Multi-Myo/grip/pinch analyser, electronic, with software

GONIOMETERS

DIMEQ Delft Instruments, DIMEQ BV, Rontgenweg 1, P.O. Box 810, 2600 AV Delft, Netherlands
Manual, in plastic (50 cm) or stainless steel (19 cm) (Moeltgen)

Biometrics Ltd, Cwmfelinfach, Gwent NP11 7HZ, U.K.
Penny & Giles electrogoniometers

HEART RATE MEASUREMENT

Beckman Instruments, Inc., 2500 Harbor Blvd, Fullerton, CA 92634, USA www.beckman.com
HR recorders

Grant Instruments (Cambridge) Ltd, Shepreth, Cambridgeshire SG8 6GB, U.K. www.grant.co.uk
Ear probe and rectal probe, etc., for use with Grant data logger (see below)

Grass Technologies, AstroMed Industrial Park, 600 East Greenwich Avenue, West Warwick, RI 02893, USA
www.grasstechnologies.com

Polar Electro Oy, Hakamantie 18, 90440 Kempele, Finland
Polar Tester, for continuous measurement, chest strap mounted

SAN-EI Instrument Co. Ltd, 223-2, Nishiokubo 2-chome, Shinjuku-ku, Tokyo, 160 Japan
Uses blood flow pulses, with fingertip pickup, finger shaft pickup, or earlobe pickup

HIGH-SPEED VIDEO

Hadland Photonics, Inc., West, 20480 Pacifica Drive, Suite D, Cupertino, CA 95014, USA

HOT WIRE ANEMOMETERS

TSI Incorporated, Environmental Measurements and Controls Division, 500 Cardigan Road, Shoreview, MN 55126, USA
www.tsi.com
VELOCICALC models

LIGHT METERS

Gossen Foto-und Lichtmesstechnik GmbH, Lina-Ammon-Str. 22, D-90471, Germany
www.gossen-photo.de
Digital luxmeters and luminance meters with PC interface and software

Tektronix Inc., P.O. Box 500, Beaverton, Oregon 97077, USA
Digital photometers

YES International Ltd, 40 High Street, Earl Shilton, Leicestershire LE9 7DG, U.K.
Eurisem meters

O_2/CO_2 BREATH GAS ANALYSERS

Beckman Instruments, Inc., 2500 Harbor Blvd, Fullerton CA 92634, USA
www.beckman.com
Grass Technologies, AstroMed Industrial Park, 600 East Greenwich Avenue, West Warwick, RI 02893, USA
www.grasstechnologies.com
Polygraph with variety of amplifiers

Medical Graphics, Birmingham, U.K.
CPX-D Cardiopulmonary Exercise Testing System

Mijnhardt B.V., Singel 45, Odijk, Netherlands
Ergo-Analyser

RATING FILMS AND VIDEOS

Tampa Manufacturing Institute, E. B. Watmough Dir., Shell Point Building, 6300 Flotilla Drive, Holmes Beach, FL 33510, USA

Van Goubergen P&M, Burgemeester Somerslaan 5, B-2350 Vosselaar, Belgium www.vangoubergen.com/videos.html
Many training films and videos, consulting, etc.

SMOKE GENERATOR

Dragerwerk AG, Moislinger Allee 53/55, Lubeck, Germany
OR Drager Safety, Inc., 10450 Stancliff, Suite 220, Houston, TX 77099, USA
Drager air flow tester

STERILISER FLUID AND TANK (FOR USE WITH O_2/CO_2 ANALYSER EQUIPMENT)

Bowak Ltd, 18-20 Sterling Way, Tilehurst, Reading RG30 6BB, U.K. www.bowak.co.uk
Milton steriliser fluid and equipment, as used for baby bottles, etc.

SOUND LEVEL METERS

Bruel & Kjer, DK-2850 Naerum, Denmark
Large variety for various applications

CEL Instruments Ltd, 35-37 Bury Mead Road, Hitchin, Herts SG5 1RT, U.K.
Instruments of various types

SPIROMETERS (FOR MEASUREMENT OF LUNG CAPACITY AND LUNG FUNCTION)

Vitalograph Inc, 8347 Quivira Road, Lenexa, Kansas 66215, USA
Vitalograph with PC interface

Mijnhardt B.V., Bunnik, Netherlands
Collins Medical, 220 Wood Road, Braintree, MA 02184-2408, USA
www.collinsmedical.com

TEMPERATURE MEASUREMENT

Airflow Development (Canada) Ltd, 1281 Matheson Boulevard, Mississauga, Ontario, Canada L4W 1R1
Botsball for combined wet bulb and globe thermometer

Casella London Ltd, Regent House, Britannia Walk, London N1 7ND
Whirling psychrometer to measure wet bulb and dry bulb at standard air velocity
Φ150 mm copper sphere globe thermometer, dry bulb and wet bulb thermometers, etc.

SCANTEC, Westkaal 7, 2170 Merksem-Antwerpen, Belgium
WiBGET: Heat stress monitor for wet bulb, dry bulb, globe temp, with PC interface

Yellow Springs Instruments, 1725 Brannum Lane, Yellow Springs, OH 45378, USA
www.ysi.com
Variety of YSI ambient-temperature-measuring instruments, rectal thermistors, etc.

TREADMILLS

Sport Engineering Ltd, 32 Stirchley Trading Estate, Hazelwell Road, Birmingham B30 2PF, U.K.
www.powerjog,co,uk
POWERJOG models

Cardiac Science Corporation, 3303 Monte Villa Parkway, Bothell, WA 98021, USA
www.cardiacscience.com
Quinton treadmills, electrocardiology, etc.

Collins Medical, 220 Wood Road, Braintree, MA 02184-2408, USA
www.collinsmedical.com

Appendices and Index

Appendix I

TABLE A.1
Critical Questioning Matrix

Description of the present way of doing things	Why this … ?	What happens to the system if it is not this … ?	What alternative is there to this … ?	What should be the … ?
What is done?	Action	Action	Action	Action
Where in system is it done?	Place	Place	Place	Place
When in sequence is it done?	Point	Point	Point	Point
Who or what does it?	Person or thing	Person or thing	Person or thing	Person or thing
How is it done?	Method	Method	Method	Method

Source: Adapted from Konz, S. and Johnson, S.L., 2008, *Work Design: Occupational Ergonomics*, 6th ed., Holcomb Hathaway. With permission.

Appendix II

TABLE A.2
Work Design Check-Sheet (clearly some cells below are not feasible)

With regard to the following aspects:	To improve the design, what can be ...			
	Eliminated?	Combined?	Simplified?	Changed?
Sequence of operations or movements				
Equipment design				
Tools				
Material(s)				
Product design				
Limb(s) used				
Muscle(s) loaded				
Vision needs				
Positions of equipment controls				
Height of equipment controls or bench				
Duration of operation or hold				
Worker Posture and variety				
Directions of actions				
Distances of reaches				
Other (state)				

OR can it be improved by adding:

Drop delivery?
Quick-action clamps?
Sliding grasp action?
Double-action effect (e.g. with the clamps)?
A fixture or jig to hold the parts?
Continuous motion instead of jerky?
Air or hydraulically operated vise?
Help of momentum?
Simultaneous work?
Gravity feed?
Sit or stand option?
Ejectors?

Appendix III

CORLETT'S PRINCIPLES FOR WORKPLACE DESIGN

N.B. These are in descending order of importance.

1. The worker should be able to maintain an upright and forward-facing posture during work.
2. Where vision is a requirement of the task, the necessary work points must be adequately visible with the head and trunk upright or with just the head inclined slightly forward.
3. All work activities should permit the worker to adopt several different, but equally healthy and safe, postures without reducing capability to do the work.
4. Work should be arranged so that it may be done at the worker's choice in either a seated or standing position. When seated the worker should be able to use the back rest of the chair, at will, without necessitating a change of movements.
5. The weight of the body, when standing, should be carried equally on both feet, and foot pedals designed accordingly.
6. Work should not be performed consistently at or above the level of the heart; even the occasional performance where force is exerted above heart level should be avoided. Where light hand work must be performed above heart level, rests for the upper arms are a requirement.
7. Work activities should be performed with the joints at about the mid-point of their range of movement. This applies particularly to the head, trunk, and upper limbs.
8. Where muscular force has to be exerted it should be by the largest appropriate muscle groups available and in a direction co-linear with the limbs concerned.
9. Where a force has to be exerted repeatedly, it should be possible to exert it with either of the arms, or either of the legs, without adjustment to the equipment.
10. Momentum should be employed to assist the worker wherever possible, and it should be reduced to a minimum if it must be overcome by muscular effort.
11. Continuous curved motions are preferable to straight-line motions involving sudden and sharp changes in direction.
12. Ballistic movements are faster, easier, and more accurate than restricted or "controlled" movements.
13. Both hands should preferably begin their micromotions (therbligs) simultaneously and finish at the same instant.
14. Both hands should not be idle at the same instant, except during rest periods.
15. Motion of arms should be in opposite and symmetrical directions, instead of in the same direction, and should be made simultaneously.

16. To reduce fatigue, motions should be confined to the lowest possible classification as listed below, the least tiring and most economical being shown first:
    1st Finger motions
    2nd Finger and wrist motions
    3rd Finger, wrist, and lower arm motions
    4th Finger, wrist, lower and upper arm motions
    5th Wrist, lower and upper arm, and body motions
17. Rest pauses should allow for all loads experienced at work, including environmental and information loads, and the time interval between successive rest periods.
18. Two or more tools should be combined wherever possible.
19. Gravity feed containers should be used to deliver the material as close to the point of assembly or use as possible. This delivery point should be near the height of the point of use, to eliminate any lifting or change in direction when carrying the parts.
20. Ejectors should be used to remove the finished part.
21. Use "drop delivery," whereby the operator may deliver the finished article, by releasing it in the position in which it was completed, without moving to dispose of it.
22. All materials and tools should be located within the "normal" reach work areas.
23. Consideration should always be given to the transfer of work from the hands to the feet, or other parts of the body.
24. Tools and materials should be so located as to permit a proper sequence of micromotions (therbligs). The part required at the beginning of the cycle should be next to the point of release of the finished piece from the former cycle.
25. Sequence of motions should be arranged to build rhythm and automaticity into the operation.

(From Corlett, E.N., 1978, The human body at work: new principles for designing workspaces and methods, *Management Services*, May, pp. 20–25, 52, 53; used with the agreement of the publisher).

Appendix IV

TABLE A.4
Rating Scale

Rating	Description	Walking Speed	
		km/h	mph[a]
0	No activity	0	0
50	**Very slow:** clumsy, fumbling movements; the operator appears to be half asleep, with no interest in the job	3.2	2
75	**Normal:** steady, deliberate, unhurried performance, as of an operator not paid by the piece but under proper supervision; looks slow, but time is not being wasted intentionally while under observation	4.8	3
100	**Standard:** Brisk, business-like performance, as of an average qualified operator paid by the piece; the necessary standard of quality and accuracy is achieved with confidence	6.4	4
125	**Very fast:** the operator exhibits a high degree of assurance, dexterity, and coordination of movement, well above that of an average trained operator	8.0	5
150	**Exceptionally fast:** requires intense effort and concentration, and is unlikely to be kept up for long periods; a "virtuoso" performance achieved by only a few outstanding operators	9.6	6

[a] *Although not an SI unit it is included here for reference only because these were the original speeds used.*

N.B. Rating estimates are made in multiples of 5.

This assumes that the person is of average height and physique, unladen, and walking in a straight line on a smooth and level surface without obstructions.

The scale conforms to the British standard and is known as the 0–100 scale, which has an advantage over others in that 0 represents zero activity and 100 represents the desired normal rate of working (i.e., the standard rate).

Source: Adapted from *Introduction to Work Study* (4th [revised ed.], ©1992) International Labour Organisation. With permission.

Appendix V

TABLE A.5
Fundamental Hand Motions (also called *therbligs*)

Term	Meaning
Select (St)	Take a broad view & combine with the Search aspect when it becomes: Hunt for and locate an object among several other objects
Grasp (G)	Close fingers around an object preparatory to picking it up, holding it, or manipulating it
Transport Empty (TE)	Move the hand without resistance towards, or away from, an object
Transport Loaded (TL)	Move an object from one location to another OR move an empty hand against resistance
Hold (H)	Retain an object after Grasp without movement of object taking place
Release Load (RL)	Cease holding of an object
Position (P)	Locate an object so that it is ready for the next therblig
Pre-Position (PP)	Locate an object in a predetermined place OR locate or orient it for a subsequent activity
Inspect (I)	Examine an object to determine whether or not it complies with standard size, shape, colour, or some other attribute(s)
Assemble (A)	Place an object into or on another object with which it becomes an integral part
Disassemble (DA)	Separate an object from another object of which it is an integral part
Use (U)	Manipulate a tool, device, or piece of apparatus for its intended purpose
Unavoidable Delay (UD)	A delay beyond the control of the worker
Avoidable Delay (AD)	Any delay of the worker for which he or she is responsible AND over which the worker has control
Plan (Pn)	The mental process which precedes a physical activity
Rest to overcome fatigue (R)	Fatigue or delay factor or allowance for worker to recover from fatigue

Source: Summarised from Barnes, R.M., 1980, *Motion and Time Study* (7th ed.), Wiley, New York, pp. 117–120. With permission from John Wiley and Sons.

Appendix VI

TABLE A.6
Work Measurement Terms

Term	Meaning
Cycle	The set of tasks to be performed to complete one production piece or a set of office tasks to complete a single job
Cycle time	The time required to complete all the tasks involved in producing one piece (e.g., an assembly) or completing a single job
Element	One of several constituent parts of a cycle
Observed time	The time taken by the worker for an element or cycle when observed by the study person
Rating	The process of gauging the pace at which the element or cycle is performed relative to the observer's concept of the "standard" pace of performance
Basic time	The time it would have taken the worker to complete the element or cycle if it had been performed at the standard pace
Allowances	Amounts in percentages that are added onto the Basic Time to provide the worker with time to meet basic needs (toilet, drink, snack) and to recover from the adverse effects of doing the job (aches, pains, tiredness, fuzziness)
Standard time	The time we expect the worker to need to complete a cycle day in and day out over a long period of time without any long-term adverse effects on health. The time used for planning production quantities or department outputs, and the number of employees and/or machines needed to meet a given rate of production or office output.

Appendix VII

TABLE A.7
Latin Square Ordering

First: Construct the "Standard Latin Square" arrangement:

	Order							
Participant	**1st**	**2nd**	**3rd**	**4th**	**5th**	**6th**	**7th**	**8th**
1	A	B	C	D	E	F	G	H
2	B	C	D	E	F	G	H	A
3	C	D	E	F	G	H	A	B
4	D	E	F	G	H	A	B	C
5	E	F	G	H	A	B	C	D
6	F	G	H	A	B	C	D	E
7	G	H	A	B	C	D	E	F
8	H	A	B	C	D	E	F	G

N.B. The letters are the labels for treatments.

Second: Get three sets of **random numbers** such as these:

Rows: 5, 3, 6, 2, 4, 8, 1, 7

Columns: 1, 7, 6, 3, 2, 4, 8, 5

Treatments: 3, 5, 2, 6, 1, 8, 7, 4

Third: Permute the **rows** using the first set of random numbers, i.e., move row 5 to the first position, row 3 to the second, row 6 to the third, and so on:

	Order							
Participant	**1st**	**2nd**	**3rd**	**4th**	**5th**	**6th**	**7th**	**8th**
1	E	F	G	H	A	B	C	D
2	C	D	E	F	G	H	A	B
3	F	G	H	A	B	C	D	E
4	B	C	D	E	F	G	H	A
5	D	E	F	G	H	A	B	C
6	H	A	B	C	D	E	F	G
7	A	B	C	D	E	F	G	H
8	G	H	A	B	C	D	E	F

Fourth: Permute the **columns** using the second random numbers set,

i.e., leave column 1 as the 1st, move column 7 to the second position, column 6 to the 3rd, etc.:

	Order							
Participant	**1st**	**2nd**	**3rd**	**4th**	**5th**	**6th**	**7th**	**8th**
1	E	C	B	G	F	H	D	A
2	C	A	H	E	D	F	B	G
3	F	D	C	H	G	A	E	B
4	B	H	G	D	C	E	A	F
5	D	B	A	F	E	G	C	H
6	H	F	E	B	A	C	G	D
7	A	G	F	C	B	D	H	E
8	G	E	D	A	H	B	F	C

Fifth: Permute **letters** to treatments from last random number set.

Appendix VIII

TABLE A.8
Shneiderman's Table

Relative Capabilities of Humans and Machines	
Humans Generally Better	**Machines Generally Better**
Sense low level stimuli	Sense stimuli outside human's range
Detect stimuli in a noisy background	Count or measure physical quantities
Recognise constant patterns in varying situations	
	Store quantities of coded information accurately
Sense unusual and unexpected events	Monitor pre-specified events, especially infrequent
	Make rapid and consistent responses to input signals
Remember principles and strategies	
	Recall quantities of detailed information accurately
Retrieve pertinent details without *a priori* connection	Process quantitative data in pre-specified ways
Draw upon experience and adapt decisions to the situation	
Select alternatives if the original approach fails	
Reason inductively: generalise from observations	
Act in unanticipated emergencies and novel situations	Perform repetitive, pre-programmed actions reliably
	Exert great, highly controlled physical force
Apply principles to solve varied problems	
Make subjective evaluations	
Develop new solutions	
Concentrate on important tasks when overload occurs	
	Perform several activities simultaneously
	Maintain operations under heavy information load
Adapt physical response to changes in the situation	Maintain performance over extended periods of time

Source: Shneiderman, Ben, Designing the User Interface: Strategies for Effective Human–Computer Interaction, p. 76, Table 2.2 Relative Capabilities of Humans and Machines, ©1987 Pearson Education, Inc. Reproduced by permission of Pearson Education, Inc. All rights reserved.

Appendix IX

TEST FOR DIFFERENCE BETWEEN TWO SLOPES

First, calculate the regression lines for each and label the slopes as b_1 and b_2 for lines 1 and 2, respectively, determined from n_1 and n_2 values. Then, perform a t-test where

$$t = (b_1 - b_2)/S_{b1-b2}$$

and conclude that the slopes are significantly different if the "t" value is greater than the table value for $\alpha/2$ level of confidence (α usually 5%) and $(n_1 + n_2 - 4)$ degrees of freedom.

$$\text{Put } V_{b1-b2} = (S_{b1-b2})^2 = \text{variance of the data for slopes}$$

Pooled variance of both sets of dependent variable data (Vy) is given by

$$V_y = [(n_1 - 2)V_{y1} + (n_2 - 2)V_{y2}]/\,(n_1 + n_2 - 4)$$

where V_{y1} and V_{y2} are the variances of the data of the dependent variable for line 1 and line 2, respectively.

Then $V_{b1-b2} = V_y\{1/[(n_1 - 1)V_{x1}] + 1/[(n_2 - 1)V_{x2}]\}$ and substitute it in the equation for *t*, where V_{x1} and V_{x2} are the variances of the data for each of the independent variables 1 and 2.

(See, for example, Crow, E.L., Davis, F.A. & Maxfield, M.W., 1960, Statistics Manual, Dover Press, Dover, Massachusetts.)

Appendix X

TABLE A.8
Inverse Normal Distribution

Probability		0.000	0.001	0.002	0.003	0.004	0.005	0.006	0.007	0.008	0.009
		From area under the left tail to: z value & ordinate value y									
0.0000	z	-8	-3.0902	-2.8782	-2.7478	-2.6521	-2.5758	-2.5121	-2.4573	-2.4089	-2.3656
	y	0.0000	0.0034	0.0063	0.0091	0.0118	0.0145	0.0170	0.0195	0.0219	0.0243
0.0100	z	-2.3263	-2.2904	-2.2571	-2.2262	-2.1973	-2.1701	-2.1444	-2.1201	-2.0969	-2.0749
	y	0.0267	0.0290	0.0312	0.0335	0.0357	0.0379	0.0400	0.0422	0.0443	0.0464
0.0200	z	-2.0537	-2.0335	-2.0141	-1.9954	-1.9774	-1.9600	-1.9431	-1.9268	-1.9110	-1.8957
	y	0.0484	0.0505	0.0525	0.0545	0.0565	0.0584	0.0604	0.0623	0.0643	0.0662
0.0300	z	-1.8808	-1.8663	-1.8522	-1.8384	-1.8250	-1.8119	-1.7991	-1.7866	-1.7744	-1.7624
	y	0.0680	0.0699	0.0718	0.0736	0.0755	0.0773	0.0791	0.0809	0.0826	0.0844
0.0400	z	-1.7507	-1.7392	-1.7279	-1.7169	-1.7060	-1.6954	-1.6849	-1.6747	-1.6646	-1.6546
	y	0.0862	0.0879	0.0897	0.0914	0.0931	0.0948	0.0965	0.0982	0.0998	0.1015
0.0500	z	-1.6449	-1.6352	-1.6258	-1.6164	-1.6072	-1.5982	-1.5893	-1.5805	-1.5718	-1.5632
	y	0.1031	0.1048	0.1064	0.1080	0.1096	0.1112	0.1128	0.1144	0.1160	0.1176
0.0600	z	-1.5548	-1.5464	-1.5382	-1.5301	-1.5220	-1.5141	-1.5063	-1.4985	-1.4909	-1.4833
	y	0.1191	0.1207	0.1222	0.1237	0.1253	0.1268	0.1283	0.1298	0.1313	0.1328
0.0700	z	-1.4758	-1.4684	-1.4611	-1.4538	-1.4466	-1.4395	-1.4325	-1.4255	-1.4187	-1.4118
	y	0.1343	0.1357	0.1372	0.1387	0.1401	0.1416	0.1430	0.1444	0.1458	0.1473
0.0800	z	-1.4051	-1.3984	-1.3917	-1.3852	-1.3787	-1.3722	-1.3658	-1.3595	-1.3532	-1.3469
	y	0.1487	0.1501	0.1515	0.1529	0.1542	0.1556	0.1570	0.1583	0.1597	0.1610
0.0900	z	-1.3408	-1.3346	-1.3285	-1.3225	-1.3165	-1.3106	-1.3047	-1.2988	-1.2930	-1.2873
	y	0.1624	0.1637	0.1651	0.1664	0.1677	0.1690	0.1703	0.1716	0.1729	0.1742

TABLE A.8 (continued)

Inverse Normal Distribution

Probability		0.000	0.001	0.002	0.003	0.004	0.005	0.006	0.007	0.008	0.009
		From area under the left tail to: z value & ordinate value y									
0.1000	z	-1.2816	-1.2759	-1.2702	-1.2646	-1.2591	-1.2536	-1.2481	-1.2426	-1.2372	-1.2319
	y	0.1755	0.1768	0.1781	0.1793	0.1806	0.1818	0.1831	0.1843	0.1856	0.1868
0.1100	z	-1.2265	-1.2212	-1.2160	-1.2107	-1.2055	-1.2004	-1.1952	-1.1901	-1.1850	-1.1800
	y	0.1880	0.1893	0.1905	0.1917	0.1929	0.1941	0.1953	0.1965	0.1977	0.1989
0.1200	z	-1.1750	-1.1700	-1.1650	-1.1601	-1.1552	-1.1503	-1.1455	-1.1407	-1.1359	-1.1311
	y	0.2000	0.2012	0.2024	0.2035	0.2047	0.2059	0.2070	0.2081	0.2093	0.2104
0.1300	z	-1.1264	-1.1217	-1.1170	-1.1123	-1.1077	-1.1031	-1.0985	-1.0939	-1.0893	-1.0848
	y	0.2115	0.2127	0.2138	0.2149	0.2160	0.2171	0.2182	0.2193	0.2204	0.2215
0.1400	z	-1.0803	-1.0758	-1.0714	-1.0669	-1.0625	-1.0581	-1.0537	-1.0494	-1.0450	-1.0407
	y	0.2226	0.2237	0.2247	0.2258	0.2269	0.2279	0.2290	0.2300	0.2311	0.2321
0.1500	z	-1.0364	-1.0322	-1.0279	-1.0237	-1.0194	-1.0152	-1.0110	-1.0069	-1.0027	-0.9986
	y	0.2332	0.2342	0.2352	0.2362	0.2373	0.2383	0.2393	0.2403	0.2413	0.2423
0.1600	z	-0.9945	-0.9904	-0.9863	-0.9822	-0.9782	-0.9741	-0.9701	-0.9661	-0.9621	-0.9581
	y	0.2433	0.2443	0.2453	0.2463	0.2473	0.2482	0.2492	0.2502	0.2511	0.2521
0.1700	z	-0.9542	-0.9502	-0.9463	-0.9424	-0.9385	-0.9346	-0.9307	-0.9269	-0.9230	-0.9192
	y	0.2531	0.2540	0.2550	0.2559	0.2568	0.2578	0.2587	0.2596	0.2606	0.2615
0.1800	z	-0.9154	-0.9116	-0.9078	-0.9040	-0.9002	-8965	-0.8927	-0.8890	-0.8853	-0.8816
	y	0.2624	0.2633	0.2642	0.2651	0.2660	0.2669	0.2678	0.2687	0.2696	0.2705
0.1900	z	-0.8779	-0.8742	-0.8705	-0.8669	-0.8633	-0.8596	-0.8560	-0.8524	-0.8488	-0.8452
	y	0.2714	0.2722	0.2731	0.2740	0.2748	0.2757	0.2766	0.2774	0.2783	0.2791

0.2000	z	-0.8416	-0.8381	-0.8345	-0.8310	-0.8274	-0.8239	-0.8204	-0.8169	-0.8134	-0.8099
	y	0.2800	0.2808	0.2816	0.2825	0.2833	0.2841	0.2849	0.2858	0.2866	0.2874
0.2100	z	-0.8064	-0.8030	-0.7995	-0.7961	-0.7926	-0.7892	-0.7858	-0.7824	-0.7790	-0.7756
	y	0.2882	0.2890	0.2898	0.2906	0.2914	0.2922	0.2930	0.2938	0.2945	0.2953
0.2200	z	-0.7722	-0.7688	-0.7655	-0.7621	-0.7588	-0.7554	-0.7521	-0.7488	-0.7454	-0.7421
	y	0.2961	0.2969	0.2976	0.2984	0.2992	0.2999	0.3007	0.3014	0.3022	0.3029
0.2300	z	-0.7388	-0.7356	-0.7323	-0.7290	-0.7257	-0.7225	-0.7192	-0.7160	-0.7128	-0.7095
	y	0.3036	0.3044	0.3051	0.3058	0.3066	0.3073	0.3080	0.3087	0.3095	0.3102
0.2400	z	-0.7063	-0.7031	-0.6999	-0.6967	-0.6935	-0.6903	-0.6871	-0.6840	-0.6808	-0.6776
	y	0.3109	0.3116	0.3123	0.3130	0.3137	0.3144	0.3151	0.3157	0.3164	0.3171
0.2500	z	-0.6745	-0.6713	-0.6682	-0.6651	-0.6620	-0.6588	-0.6557	-0.6526	-0.6495	-0.6464
	y	0.3178	0.3184	0.3191	0.3198	0.3204	0.3211	0.3218	0.3224	0.3231	0.3237
0.2600	z	-0.6433	-0.6403	-0.6372	-0.6341	-0.6311	-0.6280	-0.6250	-0.6219	-0.6189	-0.6158
	y	0.3244	0.3250	0.3256	0.3263	0.3269	0.3275	0.3282	0.3288	0.3294	0.3300
0.2700	z	-0.6128	-0.6098	-0.6068	-0.6038	-0.6008	-0.5978	-0.5948	-0.5918	-0.5888	-0.5858
	y	0.3306	0.3313	0.3319	0.3325	0.3331	0.3337	0.3343	0.3349	0.3355	0.3360
0.2800	z	-0.5828	-0.5799	-0.5769	-0.5740	-0.5710	-0.5681	-0.5651	-0.5622	-0.5592	-0.5563
	y	0.3366	0.3372	0.3378	0.3384	0.3389	0.3395	0.3401	0.3406	0.3412	0.3417
0.2900	z	-0.5534	-0.5505	-0.5476	-0.5446	-0.5417	-0.5388	-0.5359	-0.5330	-0.5302	-0.5273
	y	0.3423	0.3429	0.3434	0.3440	0.3445	0.3450	0.3456	0.3461	0.3466	0.3472

TABLE A.8 (continued)
Inverse Normal Distribution

Probability		From area under the left tail to: z value & ordinate value y									
		0.000	0.001	0.002	0.003	0.004	0.005	0.006	0.007	0.008	0.009
0.3000	z	-0.5244	-0.5215	-0.5187	-0.5158	-0.5129	-0.5101	-0.5072	-0.5044	-0.5015	-0.4987
	y	0.3477	0.3482	0.3487	0.3493	0.3498	0.3503	0.3508	0.3513	0.3518	0.3523
0.3100	z	-0.4959	-0.4930	-0.4902	-0.4874	-0.4845	-0.4817	-0.4789	-0.4761	-0.4733	-0.4705
	y	0.3528	0.3533	0.3538	0.3543	0.3548	0.3552	0.3557	0.3562	0.3567	0.3571
0.3200	z	-0.4677	-0.4649	-0.4621	-0.4593	-0.4565	-0.4538	-0.4510	-0.4482	-0.4454	-0.4427
	y	0.3576	0.3581	0.3585	0.3590	0.3595	0.3599	0.3604	0.3608	0.3613	0.3617
0.3300	z	-0.4399	-0.4372	-0.4344	-0.4316	-0.4289	-0.4261	-0.4234	-0.4207	-0.4179	-0.4152
	y	0.3621	0.3626	0.3630	0.3635	0.3639	0.3643	0.3647	0.3652	0.3656	0.3660
0.3400	z	-0.4125	-0.4097	-0.4070	-0.4043	-0.4016	-0.3989	-0.3961	-0.3934	-0.3907	-0.3880
	y	0.3664	0.3668	0.3672	0.3676	0.3680	0.3684	0.3688	0.3692	0.3696	0.3700
0.3500	z	-0.3853	-0.3826	-0.3799	-0.3772	-0.3745	-0.3719	-0.3692	-0.3665	-0.3638	-0.3611
	y	0.3704	0.3708	0.3712	0.3715	0.3719	0.3723	0.3727	0.3730	0.3734	0.3738
0.3600	z	-0.3585	-0.3558	-0.3531	-0.3505	-0.3478	-0.3451	-0.3425	-0.3398	-0.3372	-0.3345
	y	0.3741	0.3745	0.3748	0.3752	0.3755	0.3759	0.3762	0.3766	0.3769	0.3772
0.3700	z	-0.3319	-0.3292	-0.3266	-0.3239	-0.3213	-0.3186	-0.3160	-0.3134	-0.3107	-0.3081
	y	0.3776	0.3779	0.3782	0.3786	0.3789	0.3792	0.3795	0.3798	0.3801	0.3804
0.3800	z	-0.3055	-0.3029	-0.3002	-0.2976	-0.2950	-0.2924	-0.2898	-0.2871	-0.2845	-0.2819
	y	0.3808	0.3811	0.3814	0.3817	0.3820	0.3823	0.3825	0.3828	0.3831	0.3834
0.3900	z	-0.2793	-0.2767	-0.2741	-0.2715	-0.2689	-0.2663	-0.2637	-0.2611	-0.2585	-0.2559
	y	0.3837	0.3840	0.3842	0.3845	0.3848	0.3850	0.3853	0.3856	0.3858	0.3861

0.4000	z	-0.2533	-0.2508	-0.2482	-0.2456	-0.2430	-0.2404	-0.2378	-0.2353	-0.2327	-0.2301
	y	0.3863	0.3866	0.3868	0.3871	0.3873	0.3876	0.3878	0.3881	0.3883	0.3885
0.4100	z	-0.2275	-0.2250	-0.2224	-0.2198	-0.2173	-0.2147	-0.2121	-0.2096	-0.2070	-0.2045
	y	0.3887	0.3890	0.3892	0.3894	0.3896	0.3899	0.3901	0.3903	0.3905	0.3907
0.4200	z	-0.2019	-0.1993	-0.1968	-0.1942	-0.1917	-0.1891	-0.1866	-0.1840	-0.1815	-0.1789
	y	0.3909	0.3911	0.3913	0.3915	0.3917	0.3919	0.3921	0.3922	0.3924	0.3926
0.4300	z	-0.1764	-0.1738	-0.1713	-0.1687	-0.1662	-0.1637	-0.1611	-0.1586	-0.1560	-0.1535
	y	0.3928	0.3930	0.3931	0.3933	0.3935	0.3936	0.3938	0.3940	0.3941	0.3943
0.4400	z	-0.1510	-0.1484	-0.1459	-0.1434	-0.1408	-0.1383	-0.1358	-0.1332	-0.1307	-0.1282
	y	0.3944	0.3946	0.3947	0.3949	0.3950	0.3951	0.3953	0.3954	0.3955	0.3957
0.4500	z	-0.1257	-0.1231	-0.1206	-0.1181	-0.1156	-0.1130	-0.1105	-0.1080	-0.1055	-0.1030
	y	0.3958	0.3959	0.3961	0.3962	0.3963	0.3964	0.3965	0.3966	0.3967	0.3968
0.4600	z	-0.1004	-0.0979	-0.0954	-0.0929	-0.0904	-0.0878	-0.0853	-0.0828	-0.0803	-0.0778
	y	0.3969	0.3970	0.3971	0.3972	0.3973	0.3974	0.3975	0.3976	0.3977	0.3977
0.4700	z	-0.0753	-0.0728	-0.0702	-0.0677	-0.0652	-0.0627	-0.0602	-0.0577	-0.0552	-0.0527
	y	0.3978	0.3979	0.3980	0.3980	0.3981	0.3982	0.3982	0.3983	0.3983	0.3984
0.4800	z	-0.0502	-0.0476	-0.0451	-0.0426	-0.0401	-0.0376	-0.0351	-0.0326	-0.0301	-0.0276
	y	0.3984	0.3985	0.3985	0.3986	0.3986	0.3987	0.3987	0.3987	0.3988	0.3988
0.4900	z	-0.0251	-0.0226	-0.0201	-0.0175	-0.0150	-0.0125	-0.0100	-0.0075	-0.0050	-0.0025
	y	0.3988	0.3988	0.3989	0.3989	0.3989	0.3989	0.3989	0.3989	0.3989	0.3989

TABLE A.8 (continued)
Inverse Normal Distribution

		From area under the left tail to: z value & ordinate value y									
Probability		**0.000**	**0.001**	**0.002**	**0.003**	**0.004**	**0.005**	**0.006**	**0.007**	**0.008**	**0.009**
0.5000	z	0.0000	0.0025	0.0050	0.0075	0.0100	0.0125	0.0150	0.0175	0.0201	0.0226
	y	0.3989	0.3989	0.3989	0.3989	0.3989	0.3989	0.3989	0.3989	0.3989	0.3988
0.5100	z	0.0251	0.0276	0.0301	0.0326	0.0351	0.0376	0.0401	0.0426	0.0451	0.0476
	y	0.3988	0.3988	0.3988	0.3987	0.3987	0.3987	0.3986	0.3986	0.3985	0.3985
0.5200	z	0.0502	0.0527	0.0552	0.0577	0.0602	0.0627	0.0652	0.0677	0.0702	0.0728
	y	0.3984	0.3984	0.3983	0.3983	0.3982	0.3982	0.3981	0.3980	0.3980	0.3979
0.5300	z	0.0753	0.0778	0.0803	0.0828	0.0853	0.0878	0.0904	0.0929	0.0954	0.0979
	y	0.3978	0.3977	0.3977	0.3976	0.3975	0.3974	0.3973	0.3972	0.3971	0.3970
0.5400	z	0.1004	0.1030	0.1055	0.1080	0.1105	0.1130	0.1156	0.1181	0.1206	0.1231
	y	0.3969	0.3968	0.3967	0.3966	0.3965	0.3964	0.3963	0.3962	0.3961	0.3959
0.5500	z	0.1257	0.1282	0.1307	0.1332	0.1358	0.1383	0.1408	0.1434	0.1459	0.1484
	y	0.3958	0.3957	0.3955	0.3954	0.3953	0.3951	0.3950	0.3949	0.3947	0.3946
0.5600	z	0.1510	0.1535	0.1560	0.1586	0.1611	0.1637	0.1662	0.1687	0.1713	0.1738
	y	0.3944	0.3943	0.3941	0.3940	0.3938	0.3936	0.3935	0.3933	0.3931	0.3930
0.5700	z	0.1764	0.1789	0.1815	0.1840	0.1866	0.1891	0.1917	0.1942	0.1968	0.1993
	y	0.3928	0.3926	0.3924	0.3922	0.3921	0.3919	0.3917	0.3915	0.3913	0.3911
0.5800	z	0.2019	0.2045	0.2070	0.2096	0.2121	0.2147	0.2173	0.2198	0.2224	0.2250
	y	0.3909	0.3907	0.3905	0.3903	0.3901	0.3899	0.3896	0.3894	0.3892	0.3890
0.5900	z	0.2275	0.2301	0.2327	0.2353	0.2378	0.2404	0.2430	0.2456	0.2482	0.2508
	y	0.3887	0.3885	0.3883	0.3881	0.3878	0.3876	0.3873	0.3871	0.3868	0.3866

0.6000	z	0.2533	0.2559	0.2585	0.2611	0.2637	0.2663	0.2689	0.2715	0.2741	0.2767
	y	0.3863	0.3861	0.3858	0.3856	0.3853	0.3850	0.3848	0.3845	0.3842	0.3840
0.6100	z	0.2793	0.2819	0.2845	0.2871	0.2898	0.2924	0.2950	0.2976	0.3002	0.3029
	y	0.3837	0.3834	0.3831	0.3828	0.3825	0.3823	0.3820	0.3817	0.3814	0.3811
0.6200	z	0.3055	0.3081	0.3107	0.3134	0.3160	0.3186	0.3213	0.3239	0.3266	0.3292
	y	0.3808	0.3804	0.3801	0.3798	0.3795	0.3792	0.3789	0.3786	0.3782	0.3779
0.6300	z	0.3319	0.3345	0.3372	0.3398	0.3425	0.3451	0.3478	0.3505	0.3531	0.3558
	y	0.3776	0.3772	0.3769	0.3766	0.3762	0.3759	0.3755	0.3752	0.3748	0.3745
0.6400	z	0.3585	0.3611	0.3638	0.3665	0.3692	0.3719	0.3745	0.3772	0.3799	0.3826
	y	0.3741	0.3738	0.3734	0.3730	0.3727	0.3723	0.3719	0.3715	0.3712	0.3708
0.6500	z	0.3853	0.3880	0.3907	0.3934	0.3961	0.3989	0.4016	0.4043	0.4070	0.4097
	y	0.3704	0.3700	0.3696	0.3692	0.3688	0.3684	0.3680	0.3676	0.3672	0.3668
0.6600	z	0.4125	0.4152	0.4179	0.4207	0.4234	0.4261	0.4289	0.4316	0.4344	0.4372
	y	0.3664	0.3660	0.3656	0.3652	0.3647	0.3643	0.3639	0.3635	0.3630	0.3626
0.6700	z	0.4399	0.4427	0.4454	0.4482	0.4510	0.4538	0.4565	0.4593	0.4621	0.4649
	y	0.3621	0.3617	0.3613	0.3608	0.3604	0.3599	0.3595	0.3590	0.3585	0.3581
0.6800	z	0.4677	0.4705	0.4733	0.4761	0.4789	0.4817	0.4845	0.4874	0.4902	0.4930
	y	0.3576	0.3571	0.3567	0.3562	0.3557	0.3552	0.3548	0.3543	0.3538	0.3533
0.6900	z	0.4959	0.4987	0.5015	0.5044	0.5072	0.5101	0.5129	0.5158	0.5187	0.5215
	y	0.3528	0.3523	0.3518	0.3513	0.3508	0.3503	0.3498	0.3493	0.3487	0.3482

TABLE A.8 (continued)
Inverse Normal Distribution

		From area under the left tail to: z value & ordinate value y									
Probability		**0.000**	**0.001**	**0.002**	**0.003**	**0.004**	**0.005**	**0.006**	**0.007**	**0.008**	**0.009**
0.7000	z	0.5244	0.5273	0.5302	0.5330	0.5359	0.5388	0.5417	0.5446	0.5476	0.5505
	y	0.3477	0.3472	0.3466	0.3461	0.3456	0.3450	0.3445	0.3440	0.3434	0.3429
0.7100	z	0.5534	0.5563	0.5592	0.5622	0.5651	0.5681	0.5710	0.5740	0.5769	0.5799
	y	0.3423	0.3417	0.3412	0.3406	0.3401	0.3395	0.3389	0.3384	0.3378	0.3372
0.7200	z	0.5828	0.5858	0.5888	0.5918	0.5948	0.5978	0.6008	0.6038	0.6068	0.6098
	y	0.3366	0.3360	0.3355	0.3349	0.3343	0.3337	0.3331	0.3325	0.3319	0.3313
0.7300	z	0.6128	0.6158	0.6189	0.6219	0.6250	0.6280	0.6311	0.6341	0.6372	0.6403
	y	0.3306	0.3300	0.3294	0.3288	0.3282	0.3275	0.3269	0.3263	0.3256	0.3250
0.7400	z	0.6433	0.6464	0.6495	0.6526	0.6557	0.6588	0.6620	0.6651	0.6682	0.6713
	y	0.3244	0.3237	0.3231	0.3224	0.3218	0.3211	0.3204	0.3198	0.3191	0.3184
0.7500	z	0.6745	0.6776	0.6808	0.6840	0.6871	0.6903	0.6935	0.6967	0.6999	0.7031
	y	0.3178	0.3171	0.3164	0.3157	0.3151	0.3144	0.3137	0.3130	0.3123	0.3116
0.7600	z	0.7063	0.7095	0.7128	0.7160	0.7192	0.7225	0.7257	0.7290	0.7323	0.7356
	y	0.3109	0.3102	0.3095	0.3087	0.3080	0.3073	0.3066	0.3058	0.3051	0.3044
0.7700	z	0.7388	0.7421	0.7454	0.7488	0.7521	0.7554	0.7588	0.7621	0.7655	0.7688
	y	0.3036	0.3029	0.3022	0.3014	0.3007	0.2999	0.2992	0.2984	0.2976	0.2969
0.7800	z	0.7722	0.7756	0.7790	0.7824	0.7858	0.7892	0.7926	0.7961	0.7995	0.8030
	y	0.2961	0.2953	0.2945	0.2938	0.2930	0.2922	0.2914	0.2906	0.2898	0.2890
0.7900	z	0.8064	0.8099	0.8134	0.8169	0.8204	0.8239	0.8274	0.8310	0.8345	0.8381
	y	0.2882	0.2874	0.2866	0.2858	0.2849	0.2841	0.2833	0.2825	0.2816	0.2808

0.8000	z	0.8416	0.8452	0.8488	0.8524	0.8560	0.8596	0.8633	0.8669	0.8705	0.8742
	y	0.2800	0.2791	0.2783	0.2774	0.2766	0.2757	0.2748	0.2740	0.2731	0.2722
0.8100	z	0.8779	0.8816	0.8853	0.8890	0.8927	0.8965	0.9002	0.9040	0.9078	0.9116
	y	0.2714	0.2705	0.2696	0.2687	0.2678	0.2669	0.2660	0.2651	0.2642	0.2633
0.8200	z	0.9154	0.9192	0.9230	0.9269	0.9307	0.9346	0.9385	0.9424	0.9463	0.9502
	y	0.2624	0.2615	0.2606	0.2596	0.2587	0.2578	0.2568	0.2559	0.2550	0.2540
0.8300	z	0.9542	0.9581	0.9621	0.9661	0.9701	0.9741	0.9782	0.9822	0.9863	0.9904
	y	0.2531	0.2521	0.2511	0.2502	0.2492	0.2482	0.2473	0.2463	0.2453	0.2443
0.8400	z	0.9945	0.9986	1.0027	1.0069	1.0110	1.0152	1.0194	1.0237	1.0279	1.0322
	y	0.2433	0.2423	0.2413	0.2403	0.2393	0.2383	0.2373	0.2362	0.2352	0.2342
0.8500	z	1.0364	1.0407	1.0450	1.0494	1.0537	1.0581	1.0625	1.0669	1.0714	1.0758
	y	0.2332	0.2321	0.2311	0.2300	0.2290	0.2279	0.2269	0.2258	0.2247	0.2237
0.8600	z	1.0803	1.0848	1.0893	1.0939	1.0985	1.1031	1.1077	1.1123	1.1170	1.1217
	y	0.2226	0.2215	0.2204	0.2193	0.2182	0.2171	0.2160	0.2149	0.2138	0.2127
0.8700	z	1.1264	1.1311	1.1359	1.1407	1.1455	1.1503	1.1552	1.1601	1.1650	1.1700
	y	0.2115	0.2104	0.2093	0.2081	0.2070	0.2059	0.2047	0.2035	0.2024	0.2012
0.8800	z	1.1750	1.1800	1.1850	1.1901	1.1952	1.2004	1.2055	1.2107	1.2160	1.2212
	y	0.2000	0.1989	0.1977	0.1965	0.1953	0.1941	0.1929	0.1917	0.1905	0.1893
0.8900	z	1.2265	1.2319	1.2372	1.2426	1.2481	1.2536	1.2591	1.2646	1.2702	1.2759
	y	0.1880	0.1868	0.1856	0.1843	0.1831	0.1818	0.1806	0.1793	0.1781	0.1768

TABLE A.8 (continued)
Inverse Normal Distribution

Probability		From area under the left tail to: z value & ordinate value y									
		0.000	**0.001**	**0.002**	**0.003**	**0.004**	**0.005**	**0.006**	**0.007**	**0.008**	**0.009**
0.9000	z	1.2816	1.2873	1.2930	1.2988	1.3047	1.3106	1.3165	1.3225	1.3285	1.3346
	y	0.1755	0.1742	0.1729	0.1716	0.1703	0.1690	0.1677	0.1664	0.1651	0.1637
0.9100	z	1.3408	1.3469	1.3532	1.3595	1.3658	1.3722	1.3787	1.3852	1.3917	1.3984
	y	0.1624	0.1610	0.1597	0.1583	0.1570	0.1556	0.1542	0.1529	0.1515	0.1501
0.9200	z	1.4051	1.4118	1.4187	1.4255	1.4325	1.4395	1.4466	1.4538	1.4611	1.4684
	y	0.1487	0.1473	0.1458	0.1444	0.1430	0.1416	0.1401	0.1387	0.1372	0.1357
0.9300	z	1.4758	1.4833	1.4909	1.4985	1.5063	1.5141	1.5220	1.5301	1.5382	1.5464
	y	0.1343	0.1328	0.1313	0.1298	0.1283	0.1268	0.1253	0.1237	0.1222	0.1207
0.9400	z	1.5548	1.5632	1.5718	1.5805	1.5893	1.5982	1.6072	1.6164	1.6258	1.6352
	y	0.1191	0.1176	0.1160	0.1144	0.1128	0.1112	0.1096	0.1080	0.1064	0.1048
0.9500	z	1.6449	1.6546	1.6646	1.6747	1.6849	1.6954	1.7060	1.7169	1.7279	1.7392
	y	0.1031	0.1015	0.0998	0.0982	0.0965	0.0948	0.0931	0.0914	0.0897	0.0879
0.9600	z	1.7507	1.7624	1.7744	1.7866	1.7991	1.8119	1.8250	1.8384	1.8522	1.8663
	y	0.0862	0.0844	0.0826	0.0809	0.0791	0.0773	0.0755	0.0736	0.0718	0.0699
0.9700	z	1.8808	1.8957	1.9110	1.9268	1.9431	1.9600	1.9774	1.9954	2.0141	2.0335
	y	0.0680	0.0662	0.0643	0.0623	0.0604	0.0584	0.0565	0.0545	0.0525	0.0505
0.9800	z	2.0537	2.0749	2.0969	2.1201	2.1444	2.1701	2.1973	2.2262	2.2571	2.2904
	y	0.0484	0.0464	0.0443	0.0422	0.0400	0.0379	0.0357	0.0335	0.0312	0.0290
0.9900	z	2.3264	2.3656	2.4089	2.4573	2.5121	2.5758	2.6521	2.7478	2.8782	3.0903
	y	0.0267	0.0243	0.0219	0.0195	0.0170	0.0145	0.0118	0.0091	0.0063	0.0034

Index

N

O

P

R

S

9781420067361

T - #0003 - 101024 - C0 - 234/156/12 - SB - 9781420067361 - Gloss Lamination